Advances in Neuroscience in Anesthesia and Critical Care

Editor

W. ANDREW KOFKE

ANESTHESIOLOGY CLINICS

www.anesthesiology.theclinics.com

Consulting Editor
LEE A. FLEISHER

September 2016 • Volume 34 • Number 3

This work is published in collaboration with the Society for Neuroscience in Anesthesiology and Critical Care.

ELSEVIER

1600 John F. Kennedy Boulevard • Suite 1800 • Philadelphia, Pennsylvania, 19103-2899

http://www.theclinics.com

ANESTHESIOLOGY CLINICS Volume 34, Number 3
September 2016 ISSN 1932-2275, ISBN-13: 978-0-323-46250-1

Editor: Patrick Manley
Developmental Editor: Kristen Helm

Anesthesiology Clinics (ISSN 1932-2275) is published quarterly by Elsevier Inc., 360 Park Avenue South, New York, NY 10010-1710. Months of issue are March, June, September, and December. Periodicals postage paid at New York, NY and at additional mailing offices. Subscription prices are $100.00 per year (US student/resident), $330.00 per year (US individuals), $400.00 per year (Canadian individuals), $596.00 per year (US institutions), $753.00 per year (Canadian institutions), $225.00 per year (Canadian and foreign student/resident), $455.00 per year (foreign individuals), and $753.00 per year (foreign institutions). To receive student and resident rate, orders must be accompanied by name of affiliated institution, date of term, and the *signature* of program/residency coordinator on institutions letterhead. Orders will be billed at individual rate until proof of status is received. Foreign air speed delivery is included in all *Clinics'* subscription prices. All prices are subject to change without notice. POSTMASTER: Send address changes to *Anesthesiology Clinics*, Elsevier Health Sciences Division, Subscription Customer Service, 3251 Riverport Lane, Maryland Heights, MO 63043. Customer Service (orders, claims, online, change of address): Elsevier Health Sciences Division, Subscription Customer Service, 3251 Riverport Lane, Maryland Heights, MO 63043. **Tel:1-800-654-2452 (U.S. and Canada); 314-447-8871 (outside U.S. and Canada). Fax: 314-447-8029. E-mail: journalscustomerservice-usa@elsevier.com (for print support); journalsonlinesupport-usa@elsevier.com (for online support).**

Reprints. For copies of 100 or more of articles in this publication, please contact the Commercial Reprints Department, Elsevier Inc., 360 Park Avenue South, New York, NY 10010-1710. Tel.: 212-633-3874; Fax: 212-633-3820; E-mail: reprints@elsevier.com.

Anesthesiology Clinics, is also published in Spanish by McGraw-Hill Inter-americana Editores S. A., P.O. Box 5-237, 06500 Mexico D. F., Mexico.

Anesthesiology Clinics, is covered in *MEDLINE/PubMed (Index Medicus), Current Contents/Clinical Medicine, Excerpta Medica, ISI/BIOMED*, and *Chemical Abstracts*.

Contributors

CONSULTING EDITOR

LEE A. FLEISHER, MD, FACC, FAHA
Robert D. Dripps Professor and Chair of Anesthesiology and Critical Care, Professor of Medicine, Perelman School of Medicine, University of Pennsylvania, Philadelphia, Pennsylvania

EDITOR

W. ANDREW KOFKE, MD, MBA
Professor; Director Neuroscience in Anesthesiology and Critical Care Program; Co-Director, Neurocritical Care; Co-Director, Perioperative Medicine and Pain Clinical Research Unit, Department of Anesthesiology and Critical Care, University of Pennsylvania, Philadelphia, Pennsylvania

SECTION EDTIORS

WILLIAM M. ARMSTEAD, PhD
Research Professor, Departments of Anesthesiology and Critical Care; Pharmacology, University of Pennsylvania, Philadelphia, Pennsylvania

DHANESH K. GUPTA, MD
Professor of Anesthesiology, Chief of the Division of Neuroanesthesiology, Chief of Off-Site Anesthesiology, Chief of the Division of Otolaryngology—Head & Neck Surgery Anesthesiology, Chief Safety Officer for Intraoperative MRI, Neuroanesthesiology Co-Lead, Duke Health Spine Co-Management Program, Department of Anesthesiology, Duke University Medical Center, Durham, North Carolina

MARTIN SMITH, MBBS, FRCA, FFICM
Consultant and Honorary Professor in Neuroanaesthesia and Neurocritical Care, National Institute for Health Research UCLH Biomedical Research Centre, Department of Neuroanaesthesia and Neurocritical Care, The National Hospital for Neurology and Neurosurgery, University College London Hospitals, London, United Kingdom

MICHAEL L. "LUKE" JAMES, MD, FAHA, FNCS
Associate Professor, Divisions of Neuroanesthesiology and Critical Care Medicine, Department of Anesthesiology; Division of Neurocritical Care, Department of Neurology, Duke University, Durham, North Carolina

AUTHORS

WILLIAM M. ARMSTEAD, PhD
Research Professor, Departments of Anesthesiology and Critical Care; Pharmacology, University of Pennsylvania, Philadelphia, Pennsylvania

RAFI AVITSIAN, MD
Vice Chair for Professional Development; Associate Professor of Anesthesiology, Department of General Anesthesiology, Cleveland Clinic, Cleveland, Ohio

M. DUSTIN BOONE, MD
Department of Anesthesia, Beth Israel Deaconess Medical Center, Harvard Medical School, Boston, Massachusetts

VERONICA CRESPO, MD
Fellow, Department of Anesthesiology, Duke University, Durham, North Carolina

JEREMY S. DORITY, MD
Assistant Professor, Department of Anesthesiology, University of Kentucky College of Medicine, Lexington, Kentucky

SAMUEL GRODOFSKY, MD
Department of Anesthesiology and Critical Care, Hospital of the University of Pennsylvania, Philadelphia, Pennsylvania

PACO S. HERSON, PhD
Professor, Departments of Anesthesiology and Pharmacology, University of Colorado Denver, Aurora, Colorado

MICHAEL L. "LUKE" JAMES, MD, FAHA, FNCS
Associate Professor, Divisions of Neuroanesthesiology and Critical Care Medicine, Department of Anesthesiology; Division of Neurocritical Care, Department of Neurology, Duke University, Durham, North Carolina

VESNA JEVTOVIC-TODOROVIC, MD, PhD, MBA
Professor; Chair, Department of Anesthesiology, University of Colorado School of Medicine, Aurora, Colorado

SAYURI JINADASA, MD
Departments of General Surgery and Anesthesia, Beth Israel Deaconess Medical Center, Harvard Medical School, Boston, Massachusetts

MATTHEW A. KIRKMAN, MBBS, MRCS, MEd
Neurocritical Care Unit, The National Hospital for Neurology and Neurosurgery, University College London Hospitals, London, United Kingdom

ANTOUN KOHT, MD
Professor of Anesthesiology, Neurological Surgery and Neurology; Chief Neurosurgical Anesthesia and Neurosurgical Anesthesia Fellowship Director, Northwestern University, Feinberg School of Medicine, Chicago, Illinois

SANDRA B. MACHADO, MD
Staff Anesthesiologist; Assistant Professor of Anesthesiology, Department of General Anesthesiology, Cleveland Clinic, Cleveland, Ohio

JEFFREY S. OLDHAM, MD
Assistant Professor, Department of Anesthesiology, University of Kentucky College of Medicine, Lexington, Kentucky

NIDIA QUILLINAN, PhD
Assistant Professor, Department of Anesthesiology, University of Colorado Denver, Aurora, Colorado

TOD B. SLOAN, MD, MBA, PhD
Professor Emeritus, Department of Anesthesiology, University of Colorado School of Medicine, Aurora, Colorado

MARTIN SMITH, MBBS, FRCA, FFICM
Consultant and Honorary Professor in Neuroanaesthesia and Neurocritical Care, National Institute for Health Research UCLH Biomedical Research Centre, Department of Neuroanaesthesia and Neurocritical Care, The National Hospital for Neurology and Neurosurgery, University College London Hospitals, London, United Kingdom

RICHARD J. TRAYSTMAN, PhD
Distinguished University Professor; Vice Chancellor for Research, Departments of Pharmacology, Anesthesiology, Emergency Medicine and Neurology, University of Colorado Denver, Aurora, Colorado

Contents

Basic Neuroscience
William M. Armstead

Over a decade ago, alarming findings were reported that exposure of the
very young and very old animals to clinically used general anesthetics
could be detrimental to their brains. The evidence presented suggested
that the exposure to commonly used gaseous and intravenous general anesthetics induces the biochemical and morphologic changes in the immature and aging neurons ultimately resulting in their demise. More alarming
was the demonstration of significant cognitive and behavioral impairments
noted long after the initial anesthesia exposure. This article provides an
overview of anesthesia-induced developmental neurotoxicity and com-
mentary on the effects of general anesthesia on the aging brain.

Every year in the United States, millions of individuals incur ischemic brain
injury from stroke, cardiac arrest, or traumatic brain injury. These acquired
brain injuries can lead to death or long-term neurologic and neuropsychological impairments. The mechanisms of ischemic and traumatic brain
injury that lead to these deficiencies result from a complex interplay of
interdependent molecular pathways, including excitotoxicity, acidotoxicity, ionic imbalance, oxidative stress, inflammation, and apoptosis. This
article reviews several mechanisms of brain injury and discusses recent
developments. Although much is known from animal models of injury, it
has been difficult to translate these effects to humans.

This article provides a review of cerebral autoregulation, particularly as it
relates to the clinician scientist experienced in neuroscience in anesthesia

and critical care. Topics covered are biological mechanisms; methods used for assessment of autoregulation; effects of anesthetics; role in control of cerebral hemodynamics in health and disease; and emerging areas, such as role of age and sex in contribution to dysautoregulation. Emphasis is placed on bidirectional translational research wherein the clinical informs the study design of basic science studies, which, in turn, informs the clinical to result in development of improved therapies for treatment of central nervous system conditions.

Anesthesia for Neurosurgery and Interventional Radiology
Dhanesh K. Gupta

Samuel Grodofsky

This review includes a summary of contemporary theories of pain processing and advocates a multimodal analgesia approach for providing perioperative care. A summary of various medication classes and anesthetic techniques is provided that highlights evidence emerging from neurosurgical literature. This summary covers opioid management, acetaminophen, nonsteroidal antiinflammatories, ketamine, lidocaine, dexmedetomidine, corticosteroids, gabapentin, and regional anesthesia for neurosurgery. At present, there is not enough investigation into these areas to describe best practices for treating or preventing chronic pain in neurosurgery; but providers can identify a wider range of options available to personalize perioperative care strategies.

Rafi Avitsian and Sandra B. Machado

Involvement of the Anesthesiologist in the early stages of care for acute ischemic stroke patient undergoing endovascular treatment is essential. Anesthetic management includes the anesthetic technique (general anesthesia vs sedation), a matter of much debate and an area in need of well-designed prospective studies. The large numbers of confounding factors make the design of such studies a difficult process. A universally agreed point in the endovascular management of acute ischemic stroke is the importance of decreasing the time to revascularization. Hemodynamic and ventilatory management and implementation of neuroprotective modalities and treatment of acute procedural complications are important components of the anesthetic plan.

Neuromonitoring
Martin Smith

Matthew A. Kirkman and Martin Smith

The monitoring of systemic and central nervous system physiology is central to the management of patients with neurologic disease in the perioperative and critical care settings. There exists a range of invasive and noninvasive and global and regional monitors of cerebral hemodynamics, oxygenation, metabolism, and electrophysiology that can be used to guide treatment decisions after acute brain injury. With mounting evidence that a single neuromonitor cannot comprehensively detect all instances of cerebral compromise, multimodal neuromonitoring allows an individualized approach to patient management based on monitored physiologic variables rather than a generic one-size-fits-all approach targeting predetermined and often empirical thresholds.

Advances in electrophysiological monitoring have improved the ability of surgeons to make decisions and minimize the risks of complications during surgery and interventional procedures when the central nervous system (CNS) is at risk. Individual techniques have become important for identifying or mapping the location and pathway of critical neural structures. These techniques are also used to monitor the progress of procedures to augment surgical and physiologic management so as to reduce the risk of CNS injury. Advances in motor evoked potentials have facilitated mapping and monitoring of the motor tracts in newer, more complex procedures.

A mismatch between cerebral oxygen supply and demand can lead to cerebral hypoxia/ischemia and deleterious outcomes. Cerebral oxygenation monitoring is an important aspect of multimodality neuromonitoring. It is increasingly deployed whenever intracranial pressure monitoring is indicated. Although there is a large body of evidence demonstrating an association between cerebral hypoxia/ischemia and poor outcomes, it remains to be determined whether restoring cerebral oxygenation leads to improved outcomes. Randomized prospective studies are required to address uncertainties about cerebral oxygenation monitoring and management. This article describes the different methods of monitoring cerebral oxygenation, their indications, evidence base, limitations, and future perspectives.

Neurocritical Care
Michael L. "Luke" James

Traumatic brain injury (TBI) is a physical insult (a bump, jolt, or blow) to the brain that results in temporary or permanent impairment of normal brain function. TBI describes a heterogeneous group of disorders. The resulting secondary injury, namely brain swelling and its sequelae, is the reason why patients with these vastly different initial insults are homogenously treated. Much of the evidence for the management of TBI is poor or conflicting, and thus definitive guidelines are largely unavailable for clinicians at this time. A substantial portion of this article focuses on discussing the controversies in the management of TBI.

ANESTHESIOLOGY CLINICS

FORTHCOMING ISSUES

December 2016
Medically Complex Patients
Robert B. Schonberger and
Stanley H. Rosenbaum, *Editors*

March 2017
Obstetric Anesthesia
Robert R.Gaiser and Onyi Onuoha,
Editors

June 2017
Pharmacology
Alan D. Kaye, *Editor*

RECENT ISSUES

June 2016
Pain Management
Perry G. Fine and Michael A. Ashburn,
Editors

March 2016
Preoperative Evaluation
Debra Domino Pulley and
Deborah C. Richman, *Editors*

December 2015
Value-Based Care
Lee A. Fleisher, *Editor*

RELATED INTEREST

Neurologic Clinics, August 2016 (Vol. 34, No. 3)
Case Studies in Neurology
Randolph W. Evans, *Editor*
Available at: http://www.neurologic.theclinics.com/

Foreword

Anesthesiologists as Clinical Neuroscientists—One Future for Our Specialty

Lee A. Fleisher, MD, FACC, FAHA
Consulting Editor

When you ask a candidate for an anesthesiology residency why they chose the field, they frequently cite their interest in clinical pharmacology and physiology. However, when you consider the mechanism of action of the agents we utilize to achieve anesthesia, the title of clinical neuroscientist should also be included. The neuroscience underlying the mechanisms of anesthesia, care of the neurosurgical patient, and neuromonitoring has advanced significantly over the past several decades and warrants a thorough discussion for all practicing anesthesiologists.

The Society for Neuroscience in Anesthesiology and Critical Care (SNACC) has been an active multidisciplinary organization focusing on advancing the art and science of the care of the neurologically impaired patient through education, training, and research in perioperative neuroscience. After the decision to commission an issue on this topic was made, it was clear that SNACC was the ideal partner. W. Andrew Kofke, MD, MBA, the current President of SNACC, accepted the position of Editor of the issue. He is Professor and Director Neuroscience in Anesthesiology and Critical Care Program and Co-Director of Neurocritical Care at the Perelman School of Medicine of the University of Pennsylvania. He is also an NIH-funded investigator in the area of neuroscience. He has commissioned an outstanding group of fellow Board

---

This work is published in collaboration with the Society for Neuroscience in Anesthesiology and Critical Care.

Anesthesiology Clin 34 (2016) xiii–xiv
http://dx.doi.org/10.1016/j.anclin.2016.06.014
1932-2275/16/$ – see front matter © 2016 Published by Elsevier Inc.

members of SNACC as the Section Editors for the issue. The issue provides a guide to the current state-of-the-art, science, and care around the complex issues of neuroanesthesia and critical care.

Lee A. Fleisher, MD, FACC, FAHA
Perelman School of Medicine
University of Pennsylvania
3400 Spruce Street, Dulles 680
Philadelphia, PA 19104, USA

E-mail address:
Lee.Fleisher@uphs.upenn.edu

Preface

Update in Neuroanesthesia—An *Anesthesiology Clinics* Issue Affiliated with SNACC

W. Andrew Kofke, MD, MBA	William M. Armstead, PhD	Dhanesh K. Gupta, MD	Martin Smith, MBBS, FRCA, FFICM	Michael L. "Luke" James, MD, FAHA, FNCS

Editor

Neuroanesthesiology has a long history as the essence of the practice of anesthesiology, altering the brain to allow surgery and supporting systemic physiology to support the brain. As such, the field has evolved significantly from the era of finger on the pulse and giving urea for intraoperative brain edema. We have dealt with many historic issues, which have included induced hypotension, venous air embolism, circulatory arrest, neuroprotection, measurement and maintenance of cerebral and spinal cord blood flow, neural mechanisms of anesthesia, pain management, and neurotoxicity. Many of these transitions have been well reviewed in two historical pieces on the history of Society for Neuroscience in Anesthesiology and Critical Care (SNACC), one by Albin at 25 years[1] and one by Kofke at 40 years of SNACC.[2] This history indicates that the field is serially changing as well, demonstrated by a series of annual reviews by Pasternak and Lanier,[3–13] such that periodic reviews such as this issue of *Anesthesiology Clinics* are needed and welcomed.

This issue of *Anesthesiology Clinics* provides a timely update of several topics relevant to Neuroanesthesiology. Notably, all of the authors are members of the SNACC, and all of the editors are present or past members of the SNACC board of directors. Although the issue has not undergone review for formal endorsement by SNACC, it is nonetheless justifiably considered to be SNACC affiliated with approval of the SNACC board of directors for this designation.

This work is published in collaboration with the Society for Neuroscience in Anesthesiology and Critical Care.

Anesthesiology Clin 34 (2016) xv–xvii
http://dx.doi.org/10.1016/j.anclin.2016.06.013
1932-2275/16/$ – see front matter © 2016 Published by Elsevier Inc.

In this issue, we review recent advances in the broad areas of basic neuroscience, clinical anesthesia practice, intraoperative and critical care monitoring, and aspects of neurocritical care. Within these broad areas, readers will find topics addressing anesthetic action, neuropathophysiology, cerebral blood flow, chronic pain, anesthesia for endovascular management of stroke, multimodality monitoring, brain oxygen monitoring, traumatic brain injury, subarachnoid hemorrhage, and neuromuscular diseases.

Although not a fully comprehensive review of neuroanesthesiology, this issue of *Anesthesiology Clinics* should provide an authoritative resource on some of the most important aspects of the major areas that comprise and define neuroanesthesiology: neuroscience, clinical anesthesia for neurosurgery and neuroradiology, neuromonitoring, and neurocritical care.

W. Andrew Kofke, MD, MBA
Department of Anesthesiology and Critical Care
University of Pennsylvania
7 Dulles Building
3400 Spruce Street
Philadelphia, PA 19104-4283, USA

William M. Armstead, PhD
Department of Anesthesiology and Critical Care
University of Pennsylvania
3620 Hamilton Walk, JM3
Philadelphia, PA 19104, USA

Dhanesh K. Gupta, MD
Department of Anesthesiology
Duke University Medical Center
DUMC 3094
Durham, NC 27710, USA

Martin Smith, MBBS, FRCA, FFICM
Department of Neuroanaesthesia and
Neurocritical Care
The National Hospital for Neurology and Neurosurgery
University College London Hospitals
Queen Square
London WC1N 3BG, UK

Michael L. "Luke" James, MD, FAHA, FNCS
Divisions of Neuroanesthesiology
and Critical Care Medicine
Department of Anesthesiology
Division of Neurocritical Care
Department of Neurology
Duke University
DUMC-3094
Durham, NC 27710, USA

E-mail addresses:
kofkea@uphs.upenn.edu (W.A. Kofke)
armsteaw@uphs.upenn.edu (W.M. Armstead)

dhanesh.gupta@duke.edu (D.K. Gupta)
martin.smith@uclh.nhs.uk (M. Smith)
michael.james@duke.edu (M.L. "Luke" James)

REFERENCES

1. Albin MS, Albin MS. Celebrating silver: the genesis of a neuroanesthesiology society. NAS–>SNANSC–>SNACC. Neuroanesthesia Society. Society of Neurosurgical Anesthesia and Neurological Supportive Care. Society of Neurosurgical Anesthesia and Critical Care. J Neurosurg Anesthesiol 1997;9:296–307.
2. Kofke WA. Celebrating ruby: 40 years of NAS→SNANSC→SNACC→SNACC. J Neurosurg Anesthesiol 2012;24:260–80.
3. Pasternak JJ, Lanier WL. Neuroanesthesiology update. J Neurosurg Anesthesiol 2016;28:93–122.
4. Pasternak JJ, Lanier WL. Neuroanesthesiology update. J Neurosurg Anesthesiol 2015;27:87–122.
5. Pasternak JJ, Lanier WL. Neuroanesthesiology update. J Neurosurg Anesthesiol 2014;26:109–54.
6. Pasternak JJ, Lanier WL. Neuroanesthesiology update. J Neurosurg Anesthesiol 2013;25:98–134.
7. Pasternak JJ, Lanier WL. Neuroanesthesiology update. J Neurosurg Anesthesiol 2012;24:85–112.
8. Pasternak JJ, Lanier WL. Neuroanesthesiology update 2010. J Neurosurg Anesthesiol 2011;23:67–99.
9. Pasternak JJ, Lanier WL. Neuroanesthesiology update. J Neurosurg Anesthesiol 2010;22:86–109.
10. Pasternak JJ, Lanier WL. Neuroanesthesiology review—2007. J Neurosurg Anesthesiol 2008;20:78–104.
11. Pasternak JJ, Lanier WL. Neuroanesthesiology review—2006. J Neurosurg Anesthesiol 2007;19:70–92.
12. Pasternak JJ, Lanier WL. Neuroanesthesiology review—2005. J Neurosurg Anesthesiol 2006;18:93–105.
13. Pasternak JJ, Lanier WL. Neuroanesthesiology review—2004. J Neurosurg Anesthesiol 2005;17:2–8.

Basic Neuroscience:
William M. Armstead

General Anesthetics and Neurotoxicity

How Much Do We Know?

Vesna Jevtovic-Todorovic, MD, PhD, MBA

KEYWORDS

- Learning and memory • Immature brain • Aging brain • Synaptogenesis • GABA
- NMDA • Synaptic transmission

KEY POINTS

- The developing and aging brain could be vulnerable to anesthesia-induced neurotoxicity.
- An important mechanism for anesthesia-induced developmental neurotoxicity is widespread neuroapoptosis.
- An early exposure to anesthesia causes long-lasting impairments in neuronal communication and faulty formation of neuronal circuitries.
- Exposure to anesthesia during both extremes of brain age could result in long-lasting impairments in cognitive and behavioral performance in animals and potentially in humans.

INTRODUCTION

Over the past several decades, clinical anesthesiology has enjoyed an enormous growth with seemingly limitless ability to take care of patients in all age groups regardless of their health status. More than 300 million complex and very painful procedures are being performed annually. To achieve the level of unconsciousness and insensitivity to pain during various interventions, human brains are being subjected to a variety of general anesthetics, alone or in combinations. The long-term outcomes of this practice are being actively investigated and scrutinized in a variety of preclinical and clinical studies.

THE PHYSIOLOGY OF NEUROTRANSMITTERS AND THEIR RECEPTORS DURING BRAIN DEVELOPMENT

All key elements of neuronal development depend on a fine balance between various neurotransmitters. In particular, it has been known that 2 major neurotransmitters, glutamate and γ-aminobutyric acid (GABA), control all aspects of neuronal migration,

This work is published in collaboration with the Society for Neuroscience in Anesthesiology and Critical Care.
Department of Anesthesiology, University of Colorado School of Medicine, Mail Stop B-113, Leprino Office Building 7th Floor, 12401 East 17th Avenue, Aurora, CO 80045, USA
E-mail address: VESNA.JEVTOVIC-TODOROVIC@UCDENVER.EDU

Anesthesiology Clin 34 (2016) 439–451
http://dx.doi.org/10.1016/j.anclin.2016.04.001
1932-2275/16/$ – see front matter © 2016 Elsevier Inc. All rights reserved.

anesthesiology.theclinics.com

differentiation, maturation, and synaptogenesis, the key components of mammalian brain development.[1] Synaptogenesis involves massive dendritic branching and formation of trillions of synaptic contacts between neurons thus enabling the formation of meaningful neuronal circuitries and orderly neuronal maps. The processes by which axonal and dendritic projections find the right "target" and the most appropriate pathways for growth are very complex and not within the scope of this article. However, it is worth mentioning that the synaptogenesis and the development of neuronal processes are based on activity-dependent remodeling, suggesting that neuronal firing and interneuronal communications are crucial for timely and proper synaptogenesis.[2,3] All aspects of developmental synaptogenesis are tightly controlled by glia, which actively participate in neuron-glia signaling while providing an appropriate milieu for neuron-neuron interaction.[4] The electrical activity and synaptic signaling are strategically important during synaptogenesis, so much so that the major inhibitory neurotransmitter, GABA, serves as an excitatory neurotransmitter during early stages of synaptogenesis.[5] Although synapses are very pliable and undergo constant remodeling whereby new synapses are being formed and others are being pruned away throughout one's life, the bulk of fundamental neuronal networks and synaptic contacts are formed during developmental synaptogenesis. In humans, that time period occurs during the last trimester of in utero life and the first few years of postnatal life, while being most intense during the first several months of postnatal life.[3]

UNPHYSIOLOGIC MODULATION OF THE NEUROTRANSMITTERS AND THEIR RECEPTOR SYSTEMS DURING BRAIN DEVELOPMENT MAY RESULT IN NEUROTOXIC DAMAGE TO THE IMMATURE NEURONS

The fact that neuronal activity and communication are crucial for proper formation of synaptic contacts and for the establishment of stable receptor structures, which form the foundation for cognitive and behavioral development, brings into focus general anesthetics, a class of drugs commonly used in modern anesthesia. Their main goal is unphysiologically "switching off" or "turning down" neuronal communication for the purpose of achieving amnesia, analgesia, and hypnosis. Both gaseous (eg, nitrous oxide, isoflurane, sevoflurane, desflurane) and intravenous (eg, benzodiazepines, barbiturates, propofol, etomidate) anesthetics are frequently used for the purpose of assuring patients' comfort during painful intervention.

General anesthetics are very potent and effective in transiently inhibiting neuronal communication.[6] However, despite their widespread use, the mechanisms of their anesthetic action are not fully understood. Based on the studies published over the last few decades, it appears that there are specific cellular targets through which general anesthetics act.[7] In general, enhancement of inhibitory synaptic transmission and/or inhibition of excitatory synaptic transmission have been reported. In particular, many intravenous anesthetics, among them barbiturates, benzodiazepines, propofol, and etomidate,[7,8] as well as inhalational volatile anesthetics such as isoflurane, sevoflurane, desflurane, and halothane,[9,10] promote inhibitory neurotransmission by enhancing $GABA_A$-induced currents in neuronal tissue. For this reason, they are often referred to as GABAergic agents. On the other hand, a small number of intravenous anesthetics (eg, phencyclidine and its derivative, ketamine)[11] and inhalational anesthetics, nitrous oxide and xenon,[12,13] inhibit excitatory neurotransmission by blocking N-methyl-D-aspartate (NMDA) receptors, a subtype of glutamate receptors.

Although general anesthetics are powerful modulators of GABA and glutamate and thus cause significant imbalance in their functioning, only recently has it been recognized that general anesthetics, given in clinically relevant concentrations and combinations, are potentially damaging to neuronal cells in adult and immature brains, thus suggesting that these agents are potentially deleterious. This newly developing knowledge suggests the importance of maintaining a fine balance in neurotransmitters release and their receptor activation during critical stages of synaptogenesis.

THE NEUROTOXIC POTENTIAL OF GENERAL ANESTHETICS IN THE DEVELOPING MAMMALIAN BRAIN

More than 3 million general anesthetics are administered to pediatric patients in the United States alone.[14] In most cases, the administration of general anesthesia is the result of an exponential increase in the frequency of operating suite visits for relatively minor surgical interventions. However, heroic attempts to save very ill infants and young children have led to multiple surgeries and prolonged, deep sedations in intensive care units (ICUs), adding to the number of significant anesthesia interventions during a delicate period of human development. Also, because premature births account for more than 12% of the overall live birth rate (www.marchofdimes.com), and because an increased proportion of these premature babies, some of them as young as 20 weeks postconception, survive in neonatal ICUs, more of them undergo anesthesia on a daily basis (eg, surgical interventions, prolonged ICU sedation).

It would be overly simplistic to consider children as small adults when it comes to anesthesia management. Of particular interest for this article is that the human central nervous system (CNS) is not completely developed at birth. The brain weighs approximately 335 g at birth, doubles in size by 6 months of age, and almost triples by 12 months[2]; this is known as the brain growth spurt period,[3] or the period of synaptogenesis. During this time, trillions of synaptic connections are being formed, while each neuron is vastly expanding its dendritic surfaces to accommodate incoming axonal contacts. Because of this postnatal maturation of the CNS, the effects of general anesthetics on the immature brain deserve careful consideration.

Many years ago, it was noted that children who, during first 2 years of life, had undergone surgery and general anesthesia were experiencing a higher incidence of postoperative psychological disturbances than did older children.[15,16] It was assumed that these disturbances were a consequence of the emotional and physical trauma of surgery rather than an effect of anesthesia techniques or agents. The possibility that anesthetic agents per se might have damaging effects on the developing brain was not rigorously addressed or systematically investigated. Now, however, rapidly emerging animal and human data suggest that common general anesthetics could be detrimental to the developing brain.

During early stages of brain development, neurons have to establish their final destination and to form meaningful connections, which constitute the morphologic and physiologic bases for the formation of functional neuronal circuitries. Neurons that are not successful in making meaningful connections are considered redundant and are destined to die by programmed cell death (ie, apoptosis), which occurs naturally during normal development of the mammalian CNS. Although this is a physiologic process, apoptosis during normal development is tightly controlled, resulting in the removal of only a small percentage of neurons.[17] A disturbance of the fine balance between glutamatergic and GABAergic neurotransmission by excessive depression of neuronal activity and unphysiologic changes in the synaptic environment during a crucial stage of brain development may constitute a generic signal for developing neurons to "commit suicide." Because the goal of surgical anesthesia is to render the patient unconscious and insentient to pain, the ultimate question becomes whether general anesthetics at doses that ensure substantial receptor occupancy and profound depression of neuronal activity could be promoting excessive activation of neuroapoptosis and the death of large populations of developing neurons.

It appears[17–20] that general anesthetics do indeed cause significant and widespread neuroapoptotic degeneration of developing neurons in various mammalian species. The peak of vulnerability to anesthesia-induced neuroapoptosis in each species coincides with its peak of synaptogenesis, with much less vulnerability observed during late stages of synaptogenesis.[20,21] Aside from prominent caspase-3 staining, a hallmark of apoptotic death that could be detected at the light microscopic level, detailed examination at the ultrastructural level suggests that the initial insult, visible mainly in the nucleus, is marked by the clumping of chromatin, followed by disruption of the nuclear membrane, intermixing of the cytoplasm and nucleoplasm, and the formation of apoptotic bodies[17] (**Fig. 1**). In recent years, considerable effort has been put forth into elucidating the mechanisms of anesthesia-induced developmental neuroapoptosis, and it was found to involve several cascades of cellular events that ultimately led to neuronal deletion and impairment of proper synapse formation.

THE IMPORTANCE OF ANESTHESIA-INDUCED MODULATION OF NEUROTRANSMITTERS IN THE FORMATION OF NEURONAL NETWORKS

The morphologic changes described thus far represent substantial changes in neuronal structure that can be detected easily using histologic assessments. An important issue that has been brought to light on several occasions is that seemingly subtle changes that cannot be detected morphologically remain in surviving "normal" neurons after the grossly damaged neurons have been removed. Based on presently available evidence, these neurons may not be truly functional (ie, their communications may be faulty). The author first noted that an early exposure to general anesthesia causes long-term impairment in synaptic transmission in the hippocampus of adolescent rats (postnatal day 27–33) exposed to anesthesia at the peak of their synaptogenesis (postnatal day 7).[17] In particular, long-term potentiation was impaired significantly despite the presence of robust short-term potentiation. This observation suggested a long-lasting disturbance in neuronal circuitries in the young hippocampus, a brain region that is crucial for proper learning and memory development. A deficit in long-term potentiation was confirmed when synaptic transmission was examined using patch-clamp recordings of evoked inhibitory postsynaptic current and evoked excitatory postsynaptic current by recording from the pyramidal layer of control and anesthesia-treated rat subiculum, an important component of the hippocampal complex. Again it was noted that anesthesia-treated animals suffered from impaired synaptic transmission with inhibitory transmission affected significantly.[22]

Although the precise mechanisms responsible for the long-lasting changes in synaptic communication after anesthesia remain to be deciphered, some recent findings suggest that anesthetics impair axon targeting and inhibit axonal growth cone collapse, resulting in lack of proper response to guidance cues, thus causing errors in axon targeting[23] (**Fig. 2**).

EXCESSIVE MODULATION OF NEUROTRANSMITTERS DURING CRITICAL STAGES OF BRAIN DEVELOPMENT AND LONG-TERM EFFECTS ON ANIMAL BEHAVIOR

The above pathomorphological findings make it clear that anesthesia exposure results in neuronal deletion[20,24] and long-lasting impairment of synapses in vulnerable brain regions.[25–28] The ultimate question is whether and how these observations translate to lasting effects on behavior, and the short answer appears to be that they do. Development of cognitive abilities of animals exposed to general anesthetics at the peak of

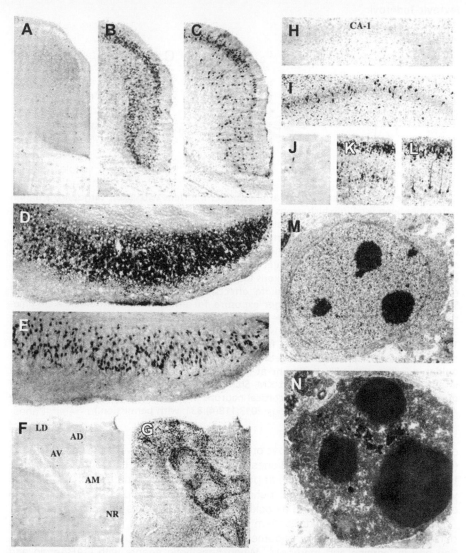

Fig. 1. Triple anesthetic cocktail induces apoptotic neurodegeneration. (A–L) Light micrographic views of various brain regions of either a control rat (A, F, H, J) or a rat exposed to the triple anesthetic cocktail (0.75-vol% isoflurane with midazolam at 9 mg/kg, subcutaneously, and nitrous oxide at 75 vol% for 6 hours) (B–E, G, I, K, L). Some sections were stained by the DeOlmos silver method (A, B, D, F, G, K); the others were immunocytochemically stained to reveal caspase-3 activation (C, E, H–J, L). The regions illustrated are the posterior cingulate/retrosplenial cortex (A–C), subiculum (D, E), anterior thalamus (F, G), rostral CA1 hippocampus (H, I), and parietal cortex (J–L). The individual nuclei shown in the anterior thalamus (F, G) are laterodorsal (LD), anterodorsal (AD), anteroventral (AV), anteromedial (AM), and nucleus reuniens (NR). (M, N) Electron micrographic scenes depicting the ultrastructural appearance of neurons undergoing apoptosis. The cell in (M) displays an early stage of apoptosis in which dense spherical chromatin balls are forming in the nucleus while the nuclear membrane remains intact; few changes are evident in the cytoplasm. The cell in (N) exhibits a much later stage of apoptosis in which the entire cell is condensed, the nuclear membrane is absent, and there is intermixing of nuclear and cytoplasmic constituents. (*From* Jevtovic-Todorovic V, Hartman RE, Izumi Y, et al. Early exposure to common anesthetic agents causes widespread neurodegeneration in the developing rat brain and persistent learning deficits. J Neurosci 2003;23(3):878; with permission.)

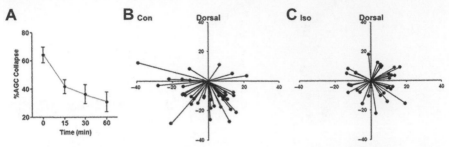

Fig. 2. To determine whether general anesthetics interfere with axon targeting, a model system has been used in which axons follow a clear, easily assessed trajectory. In this "slice overlay" assay, the axons of dissociated neocortical neurons applied to the cortical plate of an early postnatal coronal neocortex slice are directed ventrally toward the subcortical white matter by a dorsoventral gradient of Semaphorin 3A. Eighty-five percent of control axons take an appropriate ventral trajectory. In striking contrast, after 8 hours of treatment with 1.2% isoflurane, only 58% of axons take a ventral trajectory, a statistically significant reduction. Isoflurane causes immediate effects on axonal growth cone sensitivity to Semaphorin 3A but requires a longer exposure for effects on axon guidance. A time-response curve is shown depicting the percent axonal growth cone collapse to Semaphorin 3A at increasing time of exposure to isoflurane, which shows a loss of collapse in as little as 15 minutes (A). Trajectory diagrams from a slice overlay assay conducted at 5 hours, the minimum time for appropriate targeting of controls (B), shows that a disruption of axon guidance is seen with 1.8% isoflurane at this time point. Axes in (B) and (C) are measured in microns. (From Mintz CD, Barrett KM, Smith SC, et al. Anesthetics interfere with axon guidance in developing mouse neocortical neurons in vitro via a γ-aminobutyric acid type A receptor mechanism. Anesthesiology 2013;118(4):831; with permission.)

synaptogenesis lagged behind those of controls, with the gap widening into adulthood (**Fig. 3**). Even intravenous general anesthetics like propofol or thiopental in combination with ketamine (but not singly) at postnatal day 10 alters mouse behavior later in young adulthood.[29] Similar adult behavioral deficits were noted when mice were exposed at postnatal day 10 to a cocktail containing ketamine and diazepam.[30]

Although anesthesia cocktails seem to be most detrimental, ketamine given alone during early stages of brain development in rats also caused later deficits in habituation and in learning and memory.[30] When anesthetic agents with GABAergic and NMDA antagonist properties are combined, which is done frequently in the clinical setting (eg, nitrous oxide and volatile anesthetics or propofol and ketamine), cognitive deficits are more profound.[17,29,30] Although causality is difficult to establish, it is reasonable to propose that anesthesia-induced neuroapoptosis, is at least in part responsible for the observed cognitive deficits. Some recently published studies suggest that multiple, shorter-lasting exposures to anesthesia (eg, sevoflurane, ketamine) during vulnerable periods cause significant impairments in morphologic and neurocognitive development.[31,32]

General anesthesia is rarely administered in the absence of surgery and its associated pain and tissue injury. Thus, the reported neurotoxic potential of general anesthesia during brain development needs to be confirmed in the setting of surgical stimulation. Using clinically relevant concentrations of nitrous oxide and isoflurane, Shu and his colleagues[33] found that nociception enhanced neuroapoptosis and worsened long-term cognitive impairments when compared with anesthesia alone. These results would seem to undermine the hope that surgical stimulation may somehow be "protective" against anesthesia-induced neurotoxicity.

MORRIS WATER MAZE

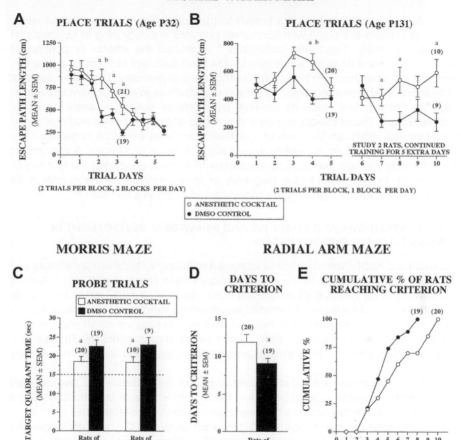

Fig. 3. Effects of neonatal triple anesthetic cocktail treatment on spatial learning. (A) Rats were tested at postnatal day 32 (P32) for their ability to learn the location of a submerged (not visible) platform. (B) Rats were retested as adults (P131) for their ability to learn a different location of the submerged platform. The graph on the left shows the path-length data from the first 5 place trials when all rats were tested. The graph on the right shows the data from rats given 5 additional training days as adults. During these trials, rats in the control group improved their performance and appeared to reach asymptotic levels, whereas the rats given the anesthetic cocktail showed no improvement. (C) Probe trial performance of rats given the anesthetic cocktail and control rats during adult testing. The dotted line represents the amount of time that animals would be expected to spend in the target quadrant in the water tank based on chance alone. (D, E) Data are shown from the radial arm maze test done on P53 to evaluate spatial working memory capabilities. (D) This histogram shows that rats given the anesthetic cocktail required significantly more days to reach a criterion demonstrating learning (8 correct responses out of the first 9 responses for 4 consecutive days) compared with controls. (E) This plot shows the days to criterion data as the cumulative percentage of rats reaching criterion in each group as a function of blocks of training days. Numbers in parentheses in each graph indicate sample sizes. [a] $P<.05$; Bonferroni corrected level: [b] $P<.005$ in (A); [b] $P<.01$ in (B). DMSO, dimethyl sulfoxide; SEM, standard error of the mean. (From Jevtovic-Todorovic V, Hartman RE, Izumi Y, et al. Early exposure to common anesthetic agents causes widespread neurodegeneration in the developing rat brain and persistent learning deficits. J Neurosci 2003;23(3):880; with permission.)

Results from rodent studies are known to translate poorly to humans, but newly emerging behavioral studies with nonhuman primates are beginning to suggest that translation is likely. Paule and colleagues[34] examined the effects of continuous neonatal (age 5 or 6 days) ketamine infusion (24 hour) sufficient to maintain a light surgical plane of anesthesia on behavioral development in primates. They observed that ketamine-treated primates exhibit long-term disturbances in all important aspects of cognitive development, such as learning, psychomotor speed, concept formation, and motivation. These effects occurred despite an absence of physiologic or metabolic trespass. Although 24 hours of anesthetic exposure would be considered unusual, it certainly occurs, especially in critically ill patients of all ages. A very recent monkey study showed that multiple exposures to sevoflurane during the first month of postnatal life resulted in higher frequency of anxiety-related behaviors later in life (at 6 months of age), suggesting the impairment of emotional behavior.[35]

GENERAL ANESTHESIA AND COGNITIVE AND BEHAVIORAL DEVELOPMENT IN HUMANS AFTER CHILDHOOD SURGERY

Backman and Kopf[36] were the first to suggest a relationship between anesthesia and long-term cognitive delay. In this report, children were anesthetized with ketamine and halothane for removal of congenital nevocytic nevi, a fairly minor procedure. Nevertheless, the investigators reported an increased incidence of cognitive impairment, described as regressive behavioral changes, lasting up to 18 months after the procedure. Children younger than 3 years were the most sensitive. These investigators voiced concern that general anesthetics may cause long-term cognitive effects.

Clinical investigations are still at an early stage, but the evidence that has emerged over the last few years suggests potentially detrimental effects of anesthetic exposure in young children on subsequent behavioral and cognitive development.

In a population-based retrospective birth-cohort study of 5357 children, Wilder and colleagues[37] found that children who received 2 or more general anesthetics before the age of 4 years were at increased risk of learning disability as adolescents (**Fig. 4**). Moreover, the risk increased with longer cumulative exposure duration (more than 2 hours). Of particular concern is their finding that the cohort exposed to

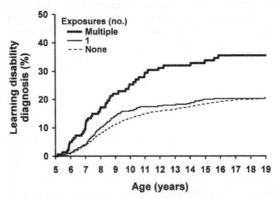

Fig. 4. Cumulative percentage of learning disabilities diagnosis by the age at exposure shown separately for those that have zero, one, or multiple anesthetic exposures before age 4 years. (*From* Wilder RT, Flick RP, Sprung J, et al. Early exposure to anesthesia and learning disabilities in a population-based birth cohort. Anesthesiology 2009;110(4):800; with permission.)

anesthesia before the age of 4 years had cognitive scores that were 2 standard deviations less than predicted, implying that early exposure to general anesthesia may have prevented achievement of full cognitive potential. In an even larger population study, Sun and colleagues[14] assessed learning disabilities in 228,961 individuals. Children who had procedures requiring anesthesia before the age of 3 years required more Medicaid services for learning disability than did children not having these procedures. Procedures in premature infants have also been associated with an excess of behavioral disabilities later in life. For example, surgically treated premature infants with patent ductus arteriosus[38] or necrotizing enterocolitis[39,40] had worse neurologic outcomes than did premature infants who were treated medically, although it may be that the severity of the disease necessitating surgical intervention contributed to the neurologic outcome.

Some newer retrospective evidence has revealed that the longer the duration of anesthesia, the lower the scores on academic achievement tests.[41] Among otherwise healthy Iowa children, the groups exposed to anesthesia at a very young age had a significantly higher percentage (about 12%–14%) performing below the fifth percentile on academic achievement tests as compared with the Iowa population as a whole (about 5%). Although the investigators caution that there may be explanations other than anesthesia exposure, this result is of concern. Sprung and colleagues[42] reported that after adjustment for comorbidities, children who underwent 2 or more exposures to anesthesia before the age of 2 years had a significantly higher risk of developing hyperactivity problems.

Collectively, this emerging clinical evidence suggests that children exposed to general anesthesia before the age of 3 years are particularly vulnerable to developing a variety of behavioral impairments. Furthermore, there appears to be a direct correlation between the duration of exposure to general anesthesia and the risk for developing cognitive impairments (ie, longer the exposure, more likely it is that the child will exhibit some form of learning disability later in life).

This rapidly emerging human evidence has to be assessed in view of the complexity and interplay between different facets of child development. Not only could the age of exposure be an important determinant of the susceptibility to anesthesia-induced developmental impairments but also the subtlety of the neurocognitive assessment could be a key factor in the ability to capture potential deficits at any given age.[43,44] For example, when children exposed to anesthesia before the age of 3 were assessed at the age of 10, the outcomes depended significantly on the type of assessment tools that were used. Although the anesthesia-exposed children did not show a significant difference in academic achievement scores as compared with unexposed children, they showed subtle but significant impairment in language abilities (in particular in receptive and expressive language) and increased risk of disability in cognition that were detected even after a single exposure to anesthesia.[44] These results suggest that the choice of assessment tools and outcome measures are crucially important in the ability to capture anesthesia-induced developmental neurotoxicity.

Although very early stages of brain development (before the age of 3 years) suggest higher vulnerability, the latest epidemiologic report suggested that children exposed to a general anesthetic between the ages of 3 and 10 years also may be vulnerable, as assessed by neurodevelopment outcomes at the age of 10, although the vulnerability may be of a different nature.[45] Using a wide range of neuropsychological tests, the investigators uncovered a higher propensity in anesthetic-exposed children to develop motor deficits, even when comorbidities or earlier injures were taken into consideration. However, the investigators concluded that there were no significant effects on language or cognitive abilities. Thus, developmental milestones may be affected in complex ways by exposure to anesthesia at different developmental stages.

Collectively, the few clinical studies to date suggest that exposure of infants and neonates to surgery and general anesthetics may cause significant neurocognitive impairment and a variety of behavioral sequelae. However, all of the studies were done retrospectively and thus could not control for the many variables that come into play during the perioperative period. The design of randomized, double-blinded prospective clinical studies of very young patients is fraught with complex issues, including ethical considerations, the lack of apoptosis biomarkers that can be used safely in a living organism, the complexity and meaningfulness of various clinical outcomes, especially neurocognitive ones, and the lack of appropriate controls.

GENERAL ANESTHESIA AND THE AGING BRAIN

Although most of this article is devoted to anesthesia-induced developmental neurotoxicity, it is important to note that the concerns regarding the effects of general anesthesia on the aging brain are being actively examined. Unlike the findings in children whereby some correlation between an early exposure to general anesthesia and lasting neurocognitive impairments have been suggested, clinical studies of postoperative cognitive decline (POCD) in elderly human population have been less conclusive. In fact, except for the initial International Study of Postoperative Cognitive Decline that has suggested that anesthetic duration could be a significant risk factor, no evidence of long-lasting cognitive influences regardless of the anesthetic agent or the anesthetic approach has been put forward. Having said that, the possibility that the patient population with ongoing but clinically undiagnosed neuropathology may prove to be the one most vulnerable to anesthesia-induced POCD should be seriously considered.

When it comes to preclinical data, the findings are interesting. Using an animal model of POCD in elderly rodents, Culley and collaborators[46] have shown that isoflurane and nitrous oxide exposure caused cognitive deficits that lasted at least a few weeks, whereas isoflurane was shown in the later study to be more of a culprit than nitrous oxide.[47] Although these findings were confirmed in other animal models where general anesthetics were examined in the absence of surgery[48] and were shown to be somewhat anesthetic dependent with isoflurane being most detrimental of all inhaled anesthetics and intravenous anesthetics being less detrimental than any of the inhaled anesthetics, POCD was concluded to be fairly transient in nature.

It is noteworthy that 2 hallmark lesions of Alzheimer disease, excessive plaque formation and tauopathy, have been shown to be influenced by general anesthetics,[49,50] thus suggesting possible interaction between anesthesia and Alzheimer pathogenesis, which may underlie POCD. The evidence regarding the role of anesthesia-induced calcium imbalance and modulation of inflammatory pathways in the development of POCD is actively being investigated.

SUMMARY

Modern neuroscience reminds us of the importance of proper functioning and fine modulation of neurotransmitters not only in aging brain but also importantly during various stages of brain development. Any degree of "tempering" with timely neurotransmitter synthesis, release, receptor activation, or inhibition could have substantial consequences on synaptogenesis, neuronal network formation, and proper cognitive and behavioral functioning.

REFERENCES

1. Komuro H, Rakic P. Modulation of neuronal migration by NMDA receptors. Science 1993;260:95–7.
2. Brown JK, Omar T, O'Regan M. Brain development and the development of tone and muscle. In: Connolly KJ, Forssberg H, editors. Neurophysiology and the neuropsychology of motor development. London: Mackeith Press; 1997. p. 1–41.
3. Dobbing J, Sands J. The brain growth spurt in various mammalian species. Early Hum Dev 1979;3:79–84.
4. Allen NJ, Barres BA. Signaling between glia and neurons: focus on synaptic plasticity. Curr Opin Neurobiol 2005;15:542–8.
5. Ben-Ari Y. Excitatory actions of GABA during development: the nature of the nurture. Nat Rev Neurosci 2002;3:728–39.
6. Hudetz AG. General anesthesia and human brain connectivity. Brain Connect 2012;2:291–302.
7. Franks NP. General anaesthesia: from molecular targets to neuronal pathways of sleep and arousal. Nat Rev Neurosci 2008;9:370–86.
8. Hirota K, Roth SH, Fujimura J, et al. GABAergic mechanisms in the action of general anesthetics. Toxicol Lett 1998;100–101:203–7.
9. Pearce RA. Effects of volatile anesthetics on GABAA receptors: Electrophysiological studies. In: Moody EJ, Skolnick P, editors. Molecular basis of anesthesia. Boca Raton (Fl): CRC Press; 2000. p. 245–72.
10. Nishikawa K, Harrison NL. The actions of sevoflurane and desflurane on the gamma-aminobutyric acid receptor type A: effects of TM2 mutations in the alpha and beta subunits. Anesthesiology 2003;99:678–84.
11. Lodge D, Anis NA. Effects of phencyclidine on excitatory amino acid activation of spinal interneurons in the cat. Eur J Pharmacol 1982;77:203–4.
12. Jevtovic-Todorovic V, Todorovic SM, Mennerick S, et al. Nitrous oxide (laughing gas) is an NMDA antagonist, neuroprotectant and neurotoxin. Nat Med 1998;4:460–3.
13. Franks NP, Dickinson R, deSousa SLM, et al. How does xenon produce anesthesia? Nature 1998;396:324.
14. Sun LS, Li G, Dimaggio C, et al. Anesthesia and neurodevelopment in children: time for an answer? Anesthesiology 2008;109:757–61.
15. Levy DM. Psychic trauma of operations in children and a note on combat neurosis. Am J Dis Child 1945;69:7–25.
16. Jackson K. Psychological preparation as a method of reducing emotional trauma of anesthesia in children. Anesthesiology 1951;12:293–300.
17. Jevtovic-Todorovic V, Hartman RE, Izumi Y, et al. Early exposure to common anesthetic agents causes widespread neurodegeneration in the developing rat brain and persistent learning deficits. J Neurosci 2003;23:876–82.
18. Loepke AW, Istaphanous GK, McAuliffe JJ 3rd, et al. The effects of neonatal isoflurane exposure in mice on brain cell viability, adult behavior, learning, and memory. Anesth Analg 2009;108:90–104.
19. Young C, Jevtovic-Todorovic V, Qin YQ, et al. Potential of ketamine and midazolam, individually or in combination, to induce apoptotic neurodegeneration in the infant mouse brain. Br J Pharmacol 2005;146:189–97.
20. Rizzi S, Carter LB, Ori C, et al. Clinical anesthesia causes permanent damage to the fetal guinea pig brain. Brain Pathol 2008;18:198–210.
21. Yon J-H, Daniel-Johnson J, Carter LB, et al. Anesthesia induces neuronal cell death in the developing rat brain via the intrinsic and extrinsic apoptotic pathways. Neuroscience 2005;35:815–27.

22. Sanchez V, Feinstein SD, Lunardi N, et al. General anesthesia causes long-term impairment of mitochondrial morphogenesis and synaptic transmission in developing rat brain. Anesthesiology 2011;115:992–1002.

23. Mintz CD, Barrett KM, Smith SC, et al. Anesthetics interfere with axon guidance in developing mouse neocortical neurons in vitro via a γ-aminobutyric acid type A receptor mechanism. Anesthesiology 2013;118:825–33.

24. Nikizad H, Yon J-H, Carter LB, et al. Early exposure to general anesthesia causes significant neuronal deletion in the developing rat brain. Ann N Y Acad Sci 2007; 1122:69–82.

25. Lunardi N, Ori C, Erisir A, et al. General anesthesia causes long-lasting disturbances in the ultrastructural properties of developing synapses in young rats. Neurotox Res 2010;17:179–88.

26. Head BP, Patel HH, Niesman IR, et al. Inhibition of p75 neurotrophin receptor attenuates isoflurane-mediated neuronal apoptosis in the neonatal central nervous system. Anesthesiology 2009;110:813–25.

27. Briner A, Nikonenko I, De Roo M, et al. Developmental stage-dependent persistent impact of propofol anesthesia on dendritic spines in the rat medial prefrontal cortex. Anesthesiology 2011;115:282–93.

28. Briner A, De Roo M, Dayer A, et al. Volatile anesthetics rapidly increase dendritic spine density in the rat medial prefrontal cortex during synaptogenesis. Anesthesiology 2010;112:546–56.

29. Fredriksson A, Pontén E, Gordh T, et al. Neonatal exposure to a combination of N-methyl-D-aspartate and gamma-aminobutyric acid type A receptor anesthetic agents potentiates apoptotic neurodegeneration and persistent behavioral deficits. Anesthesiology 2007;107:427–36.

30. Fredriksson A, Archer T, Alm H, et al. Neurofunctional deficits and potentiated apoptosis by neonatal NMDA antagonist administration. Behav Brain Res 2004; 153:367–76.

31. Han T, Hu Z, Tang YY, et al. Inhibiting Rho kinase 2 reduces memory dysfunction in adult rats exposed to sevoflurane at postnatal days 7-9. Biomed Rep 2015; 3(3):361–4.

32. Zou X, Patterson TA, Sadovova N, et al. Potential neurotoxicity of ketamine in the developing rat brain. Toxicol Sci 2009;108:149–58.

33. Shu Y, Zhou Z, Wan Y, et al. Nociceptive stimuli enhance anesthetic-induced neuroapoptosis in the rat developing brain. Neurobiol Dis 2012;45:743–50.

34. Paule MG, Li M, Allen RR, et al. Ketamine anesthesia during the first week of life can cause long-lasting cognitive deficits in rhesus monkeys. Neurotoxicol Teratol 2011;33:220–30.

35. Raper J, Alvarado MC, Murphy KL, et al. Multiple anesthetic exposure in infant monkeys alters emotional reactivity to an acute stressor. Anesthesiology 2015; 123(5):1084–92.

36. Backman ME, Kopf AW. Iatrogenic effects of general anesthesia in children: considerations in treating large congenital nevocytic nevi. J Dermatol Surg Oncol 1986;12:363–7.

37. Wilder RT, Flick RP, Sprung J, et al. Early exposure to anesthesia and learning disabilities in a population-based birth cohort. Anesthesiology 2009;110:796–804.

38. Chorne N, Leonard C, Piecuch R, et al. Patent ductus arteriosus and its treatment as risk factors for neonatal and neurodevelopmental morbidity. Pediatrics 2007; 119:1165–74.

39. Rees CM, Pierro A, Eaton S. Neurodevelopmental outcomes of neonates with medically and surgically treated necrotizing enterocolitis. Arch Dis Child Fetal Neonatal Ed 2007;92:F193–8.
40. Hintz SR, Kendrick DE, Stoll BJ, et al. Neurodevelopmental and growth outcomes of extremely low birth weight infants after necrotizing enterocolitis. Pediatrics 2005;115:696–703.
41. Block RI, Thomas JJ, Bayman EO, et al. Are anesthesia and surgery during infancy associated with altered academic performance during childhood? Anesthesiology 2012;117:494–503.
42. Sprung J, Flick RP, Katusic SK, et al. Attention-deficit/hyperactivity disorder after early exposure to procedures requiring general anesthesia. Mayo Clin Proc 2012; 87:120–9.
43. Ing C, DiMaggio C, Whitehouse A, et al. Long-term differences in language and cognitive function after childhood exposure to anesthesia. Pediatrics 2012;130: e476–85.
44. Ing CH, DiMaggio CJ, Malacova E, et al. Comparative analysis of outcome measures used in examining neurodevelopmental effects of early childhood anesthesia exposure. Anesthesiology 2014;120:1319–32.
45. Ing CH, DiMaggio CJ, Whitehouse AJ, et al. Neurodevelopmental outcomes after initial childhood anesthetic exposure between ages 3 and 10 Years. J Neurosurg Anesthesiol 2014;26(4):377–86.
46. Culley DJ, Baxter M, Yukhananov R, et al. The memory effects of general anesthesia persist for weeks in young and aged rats. Anesth Analg 2003;96:1004–9.
47. Culley DJ, Baxter MG, Crosby CA, et al. Impaired acquisition of spatial memory 2 weeks after isoflurane and isoflurane-nitrous oxide anesthesia in aged rats. Anesth Analg 2004;99:1393–7.
48. Bianchi SL, Tran T, Liu C, et al. Brain and behavior changes in 12-month-old Tg2576 and nontransgenic mice exposed to anesthetics. Neurobiol Aging 2008;29:1002–10.
49. Tang JX, Eckenhoff MF. Anesthetic effects in Alzheimer transgenic mouse models. Prog Neuropsychopharmacol Biol Psychiatry 2013;47:167–71.
50. Quiroga C, Chaparro RE, Karlnoski R, et al. Effects of repetitive exposure to anesthetics and analgesics in the Tg2576 mouse Alzheimer's model. Neurotox Res 2014;26:414–21.

Neuropathophysiology of Brain Injury

Nidia Quillinan, PhD[a], Paco S. Herson, PhD[a,b], Richard J. Traystman, PhD[a,b,c,d],*

KEYWORDS

- Mechanisms of ischemic brain injury • Excitotoxicity • Oxidative stress
- Neuroinflammation • Apoptosis • Synaptic plasticity • Neurovascular unit
- Translation

KEY POINTS

- Every year in the United States, millions of individuals incur ischemic brain injury from stroke, cardiac arrest, or traumatic brain injury. These forms of acquired brain injury can lead to death, or in many cases long-term neurologic and neuropsychological impairments.
- The mechanisms of ischemic and traumatic brain injury that lead to these deficiencies result from a complex interplay of multiple interdependent molecular pathways that include excitotoxicity, acidotoxicity, ionic imbalance, oxidative stress, inflammation, and apoptosis.
- This article briefly reviews several of the traditional, well-known mechanisms of brain injury and then discusses more recent developments and newer mechanisms.
- Although much is known concerning mechanisms of injury and the manipulation of these mechanisms to result in protection of neurons and increased behavioral performance in animal models of injury, it has been difficult to translate these effects to humans. Attention is given to why this is so and newer outcome measures of injury are discussed.

Every year in the United States, millions of individuals incur ischemic brain injury from stroke, cardiac arrest, or traumatic brain injury (TBI). These forms of acquired brain injury can lead to death, or in many cases long-term neurologic and neuropsychological impairments. The mechanisms of ischemic and traumatic brain injuries that lead to these deficiencies result from a complex interplay of multiple interdependent molecular pathways that include excitotoxicity, acidotoxicity, ionic imbalance, oxidative

---

This work is published in collaboration with the Society for Neuroscience in Anesthesiology and Critical Care.

Disclosures: No disclosures for any authors.

[a] Department of Anesthesiology, University of Colorado Denver, Anschutz Medical Campus, Aurora, CO 80045, USA; [b] Department of Pharmacology, University of Colorado Denver, Anschutz Medical Campus, Aurora, CO 80045, USA; [c] Department of Emergency Medicine, University of Colorado Denver, Anschutz Medical Campus, Aurora, CO 80045, USA; [d] Department of Neurology, University of Colorado Denver, Anschutz Medical Campus, Aurora, CO 80045, USA

* Corresponding author. Department of Pharmacology, University of Colorado Denver, Anschutz Medical Campus, 13001 East 17th Place, Building 500, Room C1000, Mail Stop 520, Aurora, CO 80045.

E-mail address: richard.traystman@ucdenver.edu

Anesthesiology Clin 34 (2016) 453–464
http://dx.doi.org/10.1016/j.anclin.2016.04.011
anesthesiology.theclinics.com

stress, inflammation, and apoptosis. This article briefly reviews several of the traditional, well-known mechanisms of brain injury and then discusses more recent developments and newer mechanisms. Although much is known concerning mechanisms of injury and the manipulation of these mechanisms to result in protection of neurons and increased behavioral performance in animal models of injury, it has been difficult to translate these effects to humans. Attention is given to why this is so and newer outcome measures of injury are discussed.

MECHANISMS OF INJURY FOLLOWING BRAIN ISCHEMIA
Excitotoxicity

The glutamate excitotoxicity hypothesis of ischemic cell damage suggests that injury is triggered by glutamate, an excitatory amino acid, released during ischemia from the intracellular compartment into the extracellular environment.[1] Glutamate is a major transmitter in the nervous system and, in addition to being required for rapid synaptic transmission for neuron-to-neuron communication, glutamate plays important roles in neuronal growth and axon guidance, brain development and maturation, and synaptic plasticity. Under normal physiologic conditions, the presence of glutamate in the synapse is regulated by active ATP-dependent transporters in neurons and glia. However, if these uptake mechanisms are impaired by metabolic disturbances brought about by ischemia, glutamate excessively accumulates, stimulating sodium (Na^+) and calcium (Ca^{2+}) fluxes into the cell through glutamate receptors, thereby injuring or killing the cell. Glutamate activates different types of ion channel–forming receptors (ionotropic) and G-protein–coupled receptors (metabotropic) that have an important role in brain function. The major ionotropic receptors activated by glutamate are commonly referred to as the N-methyl-D-aspartic acid (NMDA), alpha-amino-3-hydroxy-5-methylisoxazole-4-propionate (AMPA), and kainic acid receptors. The ionotropic receptors are ligand-gated ion channels permeable to various cations. Overactivation of these receptors leads to an increase in intracellular Ca^{2+} load and catabolic enzyme activity, which can trigger a cascade of events leading to apoptosis and necrosis. These events can include membrane depolarization, production of oxygen free radicals, and cellular toxicity. NMDA and AMPA receptor antagonists showed great promise for providing neuroprotection in animal models, but have failed to translate clinically. One complication with this approach is the unwanted side effects associated with blocking receptors that are critical for normal brain function. Alternative approaches are now being considered, such as using partial NMDA antagonists such as memantine[2] or blocking receptor interactions with postsynaptic scaffolding molecules postsynaptic density protein 93/95.[3,4] Therapies targeting molecules downstream of NMDA receptors, such as calcium-calmodulin–dependent protein kinase (CAM-KII) and death-associated protein kinase (DAPK) may also hold promise for limiting the excitotoxic cascade. In contrast, metabotropic receptors (metabotropic glutamate receptors [mGluRs]) are G-protein–coupled receptors that have been subdivided into 3 groups, based on sequence similarity, pharmacology, and intracellular signaling mechanisms. The role of mGluRs in brain injury is complex; however, most of the evidence points to a neuroprotective role, likely via antiapoptotic signaling and decreased excitability countering excitotoxicity.

Acidotoxicity

Metabolic acidosis can occur as a result of lactate accumulation during and following ischemia, or when mitochondrial respiration is dysfunctional. Acid-sensing ion channels (ASICs) represent a group of ion channels activated by protons, and act as sensors of tissue pH.[2] They belong to the epithelial sodium family of amiloride-sensitive cation channels and allow Na^+ and Ca^{2+} entry into neurons. There have been at least 6 ASIC subunits cloned and ASIC1a, ASIC2a, and ASIC2b are expressed in the brain and spinal cord. ASIC1a and ASIC2s are found in brain regions with high synaptic density and facilitate excitatory transmission. ASICs have been shown to be activated in ischemia and their activation contributes to neuronal cell death through Ca^{2+}, Na^+, and Zn^{2+} influx into the cell. Remarkably, inhibition of ASIC1a following stroke has a therapeutic window of 5 hours in experimental stroke models,[5] which is beyond that of the currently available treatment, tissue plasminogen activator (tPA).

Oxidative Stress

During normal mitochondrial respiration, cytochrome c is involved in a 4-electron transfer to reduce oxygen to water without the production of oxygen radicals.[6,7] During ischemia, when oxygen supply is limited, the electron transport chain becomes highly reduced and oxygen radicals can be produced. This process can be exacerbated by the mitochondrial Ca^{2+} accumulation that occurs during and following ischemia resulting in mitochondrial dysfunction, which can result in the formation of reactive oxygen species. Several oxygen radical species can be produced, including superoxide, perhydroxyl, hydrogen peroxide, and hydroxyl radicals. Another pathway for forming hydroxyl radical is through the reaction of superoxide and nitric oxide to form peroxynitrite. All of these reactive oxygen species, especially peroxynitrite and superoxide, can bind directly to DNA, changing its structure and causing cell injury and enhancement of apoptosis. An antioxidant/antiinflammatory agent, edaravone, is currently in use to treat acute ischemic stroke in Japan. However, this and other antioxidant compounds (NXY-059; SAINT trials[8]) have failed to improve outcome following stroke in the United States. Thus, new targeted approaches to preventing reactive oxygen species damage are needed. Targeting downstream mediators of oxidative stress are emerging therapies, such as inhibition of poly(ADP ribose) polymerase-1 or transient receptor potential melastatin-2 channels. Further research into the biological consequences following oxidative stress may lead to new directed therapies to reduce ischemic injury.

Inflammation

Inflammation plays an important role in the pathogenesis of ischemic brain injury.[9,10] The brain responds to ischemic injury with an acute and prolonged inflammatory process that is characterized by rapid activation of resident cells (microglia), production of proinflammatory mediators, and infiltration of various types of inflammatory cells, such as neutrophils, different types of T cells, macrophages, and other cells, into the ischemic brain tissue. Cytokines and chemokines contribute to ischemic brain injury and, during ischemia, cytokines such as interleukin (IL)-1, IL-6, tumor necrosis factor alpha, transforming growth factor beta, and chemokines such as cytokine-induced neutrophil chemoattractant and monocyte chemoattractant protein-1 are produced by a variety of different activated cell types, such as endothelial cells, microglia, neurons, platelets, leukocytes, and fibroblasts. The inflammatory response in brain may have various consequences on outcome, depending on the degree of inflammatory response and when it occurs. Although acute inflammatory events may be involved in secondary injury processes, more delayed inflammatory events may be reparative. However, it is difficult to elucidate the precise mechanisms of the inflammatory responses following ischemia because inflammation is a complex series of interactions between inflammatory cells and molecules, all of which could be either detrimental or beneficial. Not surprisingly, acute broad-spectrum inhibitors of inflammation reduce injury in experimental models of cerebral ischemia. The lack of success in clinical translation points to the need for a deeper understanding regarding the role of the various cellular and molecular contributors to postischemic inflammation. There is also the need to better understand the complex temporal profile of the various inflammatory mediators.

Apoptosis

Apoptosis is a genetically controlled mechanism of cell death involved in the regulation of tissue homeostasis.[11,12] Biochemical events lead to characteristic cell changes (morphology) and cell death. These changes include cell blebbing, cell shrinkage, nuclear fragmentation, chromatin condensation, chromosomal DNA fragmentation, and messenger RNA decay. Triggers of apoptosis include oxygen free radicals, death receptor ligation, DNA damage, protease activation, and ionic imbalance. The 2 major pathways of apoptosis are the extrinsic (Fas and other tumor necrosis factor receptor superfamily members and ligands) and the intrinsic (mitochondria-associated) pathways, both of which are found in the cytoplasm. The extrinsic pathway is triggered by death receptor engagement, which initiates a signaling cascade mediated by caspase-8 activation, whereas the intrinsic pathway is engaged when various apoptotic stimuli trigger the release of cytochrome c from mitochondria independently of caspase-8 activation. Both pathways ultimately cause caspase-3 activation, resulting in the degradation of cellular proteins necessary to maintain cell survival and integrity. In addition, there is a complex interplay of the Bcl-2 family of proteins, which either promote (Bax, Bak, Bad, Bim, Bid) or prevent (Bcl-2, Bcl-xL, Bcl-w) injury. Bcl-2 and its family member, Bcl-xL, are among the most powerful death-suppressing proteins and inhibit both caspase-dependent and caspase-independent cell death. Among the caspase-independent apoptotic cell death pathways is apoptosis-inducing factor (AIF). AIF is stored within the same mitochondrial compartment as cytochrome c. DNA damage (via PARP activation) and oxidative or excitotoxic stress release AIF, which is translocated to the nucleus to induce apoptosis. **Fig. 1** shows these pathways.

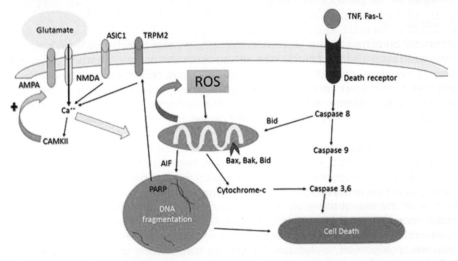

Fig. 1. Glutamate, reactive oxygen species, and apoptotic cell death pathways. Fas-L, Fas ligand; ROS, reactive oxygen species.

Translation of these injury mechanisms to neuroprotection in humans
Translation from basic science cerebral ischemia research to treatment of patients with ischemia remains a difficult challenge. Despite hundreds of compounds and interventions that provide benefit in experimental models of cerebral ischemia, efficacy in humans remains to be demonstrated. The reasons for failure to translate have been the focus of discussion for years.[13,14] Some clinicians attribute the failure to flaws in clinical trial design, others question the predictive value of current animal models, and some question the quality of preclinical data. It is likely that a combination of all these shortcomings is to blame. All of the mechanisms of ischemic injury mentioned earlier have been subject to interventions to block or inhibit the injury in preclinical animal models. Investigators have used excitatory amino acid inhibitors, ion channel blockers, free radical scavengers, antiapoptotic agents, and antiinflammatory agents with great success in ameliorating the injury from ischemia in animal models. However, none of these agents have met with success in human clinical trials. Thus, new approaches are needed that consider the complex interplay among various cell types within the brain as well as placing emphasis on examination of long-term functional recovery and plasticity.

INTERCELLULAR INTERACTIONS AND REPAIR
Neurovascular Unit

Extensive research has focused on the intracellular signaling cascades triggered by ischemic brain injury and TBI that lead to cell death. However, despite advances in the understanding of these intracellular pathways, neuroprotective strategies have failed to translate into acute treatments for ischemic stroke, cardiac arrest, or TBI. The only successful treatment to improve stroke outcome is timely revascularization (restoration of blood flow) either mechanically[15,16] or pharmacologically, with tPA.[17] A decade ago researchers began to describe the microvasculature and surrounding parenchyma in terms of multiple cellular components, including neurons, microvascular endothelial cells, astrocytes, and pericytes. The ability of astrocytes to communicate with neurons and at the same time be in close proximity to endothelial cells provided the template for the concept of the neurovascular unit.[18–20] The neurovascular unit has emerged as a convenient model to conceptualize the intercellular crosstalk in the brain regulating blood flow, integrity of the blood-brain barrier, and ultimately outcomes following injury. More recently, myelinating oligodendrocytes have been added to the picture, as cells that not only maintain myelin integrity but provide trophic support for the underlying axons and signal to surrounding neurons, astrocytes, and other components of the neurovascular unit. Thus, it is clear that emerging new research needs to focus on intercellular signaling in order to understand the mechanisms of ischemic injury in hopes of ultimately enhancing functional recovery.

Successes using revascularization strategies or therapeutic hypothermia are tempered by the short therapeutic time window, severely limiting the number of individuals eligible for such interventions. Therefore, it is imperative that new therapeutic strategies are designed to repair and restore brain function following injury. The focus here is on the role of the components of the neurovascular unit in protection and repair, independent of their role in cerebral blood flow control. Endogenous repair of injured brain involves several intertwined regenerative processes, including neurogenesis, angiogenesis, oligodendrogenesis, and synaptogenesis. In the past 2 decades, promising research has shown that stroke promotes ongoing neurogenesis in the adult and aging brain. Adult neurogenesis is the process of producing new neurons from endogenous neural stem cells residing in either the subventricular zone (SVZ) or dentate gyrus of the hippocampus. Neurogenesis following brain injury requires proliferation of endogenous stem cells, migration to the site of injury, and differentiation into mature neurons. There is evidence of injury-induced neurogenesis following various injuries, including stroke, intracranial hemorrhage, global cerebral ischemia, and TBI. However, recent data have shown limited neuronal replacement following injury, showing that neurogenesis alone is not sufficient to produce full functional recovery. Another major consideration is angiogenesis, which is the formation of new microvessels. As with neurogenesis, angiogenesis has been observed in the penumbra following ischemic stroke and emerging evidence indicates that angiogenesis enhances neurogenesis, thus indicating that these processes may be coupled and should be considered together to enhance recovery. For example, migration of neural stem cells from the SVZ has been observed to be via new microvessels.[21,22] Despite experimental strategies to enhance angiogenesis and/or neurogenesis, it seems that the injured adult brain provides an environment that is not conducive to repopulation of functional neurons. Injury causes loss of cells, which are then replaced with extracellular matrix, termed the glial or fibrotic scar. For decades, this was thought to be produced by surviving astrocytes migrating to the location of injury. However, recent work has implicated nonvascular pericytes as contributors to scar formation in the injured brain.[23,24] Recent work has shown that alteration in the composition of the scar following injury can improve the environment for repair. Thus, future research will require consideration of various cell types and their responses to injury and therapeutic interventions as a means to optimize recovery. In addition to neuronal and vascular injury, white matter injury is a major component of ischemic stroke and brain injury. Injury to myelinated axons is a consequence of oligodendrocyte disorder. In addition to demyelination, injury to oligodendrocytes also likely alters vascular integrity, as shown by recent work demonstrating oligodendrocyte-endothelial interactions in culture and during vascular remodeling.[25] Thus, new strategies to enhance cellular crosstalk during the recovery phase following brain injury are needed to enhance the repair and restoration of neuronal, vascular, and white matter function.

Neuroinflammation

In addition to oligovascular and neurovascular coupling, this entire unit interacts with the immune system. Inflammation is a major player in the outcome following all forms of brain injury. The central nervous system (CNS) inflammatory response is thought to be triggered by injured neurons. However, several recent studies have shown that all cells are susceptible to ischemic injury, with neurons being the most sensitive to direct injury, followed by oligodendrocytes, astrocytes, microglia, and endothelial cells being particularly resistant to injury. The initial immune response seems to be driven by astrocyte and microglial responses to the release of danger-associated molecular pattern (DAMP) from injured neurons. For decades it had been assumed that massive inflammatory responses drive ongoing injury. However, more recent studies have confused the issue, indicating inflammation both in injury and repair. The complex orchestrated response of the array of immune cells makes this an important and ever-changing area of research. Microglia are brain-specific macrophages that rapidly respond to injury and classic microgliosis produces a proinflammatory environment that contributes to cellular injury. Recently, an alternative or M2 activation of microglia and macrophages has been described, which produce antiinflammatory cytokines, such as IL-10, and are likely protective. It seems that strategies that enhance the alternative activation pathway reduce neuronal injury and that many cells of the immune system have subsets that have both detrimental and protective capacities. For example, brain injury causes the influx of both proinflammatory T cells (Th1, Th17) and antiinflammatory Treg cells, the balance of which ultimately contributes to the magnitude of injury and repair. Similarly, infiltrating B cells have the detrimental capacity to produce antibodies against local antigens, and also antiinflammatory cytokines such as IL-10. To further complicate the matter, emerging evidence indicates that injury-induced alterations in vascular endothelial cells play a major regulatory role in immune cell infiltration into the injured brain, particularly through blood-brain barrier dysfunction. Therefore, future studies are needed to carefully dissect the relative contributions of infiltrating and resident immune cells after injury; broad-spectrum immunotherapies are not likely to provide benefit because, in addition to blocking detrimental proinflammatory responses, treatments may block beneficial immune responses that are critical for long-term functional recovery.

FUNCTIONAL PLASTICITY

A potential issue with animal studies is an overly narrow focus on acute histologic outcome, rather than long-term functional outcomes, which are ultimately the criteria for success in humans. Pharmacologic therapies targeting histologic protection have a narrow therapeutic window, whereas neurorestorative strategies aimed at more chronic processes are likely to have a broader window for intervention. Acute brain injuries alter neuronal networks by causing neuronal cell death and alterations in excitability and synaptic contacts of surviving neurons. A better understanding of how acute brain injuries alter injured and noninjured networks could provide a means to improve function independent of neuroprotection.

Excitation/Inhibition Imbalance

A delicate balance between excitation and inhibition exists in the mammalian brain and alterations in this balance result in dysfunctional brain activity and ultimately impaired cognitive and behavioral outcomes. Changes in neuronal excitability following an acute brain injury can be the result of alterations in excitatory and/or inhibitory signaling in brain. Glutamate and gamma-aminobutyric acid (GABA) are the dominant excitatory and inhibitory neurotransmitters in the CNS, respectively. There is a growing body of literature to suggest that acute brain injury results in an imbalance between excitation and inhibition in brain. In the early stages following an acquired brain injury the balance is shifted toward excitatory transmission with increases in glutamate signaling and a downregulation of GABAergic signaling.[1,26] This imbalance can result in excitotoxicity that produces activation of cell death mechanisms. In the subacute and chronic phases following brain injury the balance shifts to favor inhibitory GABAergic signaling. GABA receptor activation can produce phasic inhibition that is rapid and mediated by synaptic GABA receptors and tonic inhibition, which is more long-lasting and associated with activation of extrasynaptic GABA receptors. Following experimental stroke there is an increase in tonic inhibition in the cortex that is evident at 1 week following injury and prevents functional recovery. When this tonic inhibition is reversed with an inhibitor of GABAA α5 signaling at delayed time points there is an improvement in functional recovery of motor and sensorimotor function that is not the result of altering histologic outcome.[27,28] Similar increases in tonic inhibition have been observed in the dentate gyrus following controlled cortical impact injury that is reversed with the GABAA α4 or δ antagonist.[29–31] Modulation of tonic inhibition may provide a novel strategy to regulate neuronal excitability and promote plasticity in networks that are altered following ischemic brain injury or TBI. Functional imaging in patients with stroke supports the hypothesis that there is overinhibition of the ipsilateral cortex, particularly from the contralateral hemisphere.[32]

An alternative to reducing inhibitory signaling to restore the excitation/inhibition balance and potentially improve functional recovery is to enhance excitatory transmission. Although this may seem counterintuitive given the excitotoxicity mechanisms that occur during the acute phase, there is evidence that increasing glutamatergic signaling in the chronic phase can be beneficial for functional plasticity and recovery. Inhibition of the glutamate transporter to enhance glutamatergic transmission in acute slices restored excitatory transmission to control levels in animals that had TBI. Positive allosteric modulators of AMPA receptor function administered in the subacute phase following stroke resulted in improved motor function through enhancement of brain-derived neurotrophic factor function.[33] Similarly, administration of D-cycloserine, which enhances NMDA receptor activation, has been shown to improve functional recovery and synaptic plasticity following injury.[34] These results suggest that the imbalance between excitation and inhibition prevents structural and synaptic plasticity in the brain and restoring the balance by antagonizing GABA receptor activity or enhancing glutamate receptor activity may serve as a promising therapy to enhance functional recovery following brain injury. This emerging area of research will lead to both new therapeutic strategies as well as a new and more complex understanding of altered signaling in the injured brain.

Synaptic Plasticity

Another potential target for therapies to improve functional recovery is synaptic plasticity, which is an experience-dependent modification of synaptic strength. Synaptic plasticity deficits are a commonly observed consequence of acute brain injury in animal models. A well-characterized form of synaptic plasticity, long-term potentiation (LTP) of CA1 pyramidal cells, is an increase in synaptic strength that is a cellular correlate of learning and memory. Impairments in hippocampal LTP have been observed in most animal models of acute brain injury including stroke, cardiac arrest, and TBI.[35-37] The loss of LTP is sustained for at least 1 month, which is well beyond the time of neuronal cell death.[38,39] This finding implies that, in addition to cell death, these forms of brain injury can produce long-lasting changes in synaptic function of surviving neurons, which likely contributes to a lack of recovery of function. The mechanisms contributing to LTP deficits remain to be elucidated, but likely candidates include excitatory/inhibitory imbalance, alterations of intracellular signaling, and prolonged neuroinflammation. LTP is initiated by activation of NMDA receptors during high-frequency stimulation and a subsequent Ca^{2+}-dependent increase in postsynaptic AMPA receptor function. Studies examining NMDA receptor function in injured brain are mixed, with reports of reduced or unchanged expression and function. Enhancing NMDA receptor function is capable of restoring hippocampal LTP following TBI.[40] Delayed administration of low doses of kainate 48 hours after ischemia also promotes the restoration of LTP in the postischemic brain.[41] Inhibition of B lymphocytes with an anti-CD20 antibody can also promote recovery of plasticity, suggesting that the loss of LTP may involve multicellular processes.[42] Targeting of synaptic plasticity deficits in the hippocampus independent of neuroprotection may serve as a promising therapy to improve cognitive function at delayed times after an acute brain injury. The hippocampus has been an important research target because of its well-established role in memory deficits following injury; however, it is likely that similar phenomena are occurring in other brain regions. Future studies are needed to assess alterations in synaptic plasticity in various brain regions not directly injured by ischemia.

Structural Plasticity

Early changes in dendritic morphology have been observed in periinfarct regions following stroke. Reduced spine density and dendritic swelling are observed during the first 24 hours following stroke onset. Although recovery of dendritic spines is observed in the subacute period following stroke, there is little evidence for this recovery following TBI.[43] This finding may have important implications for differences in therapies targeting synaptic function to improve function following different acquired brain injuries. Axonal damage and degeneration occur during the acute and subacute periods following stroke and TBI.[44,45] Recovery from stroke requires formation of new connections in injured and noninjured tissues that depends on axonal sprouting. There are many trophic and intercellular signaling processes that mediate axonal regeneration and can be targeted pharmacologically or with cell-based therapies to enhance stroke recovery.[46,47] Note that aberrant axonal sprouting can contribute to hyperexcitability and posttraumatic epilepsy.[48,49] Therefore, it is important to take care that restorative strategies enhance functionally appropriate synaptic contacts and avoid maladaptive structural plasticity and hyperexcitability. Cell-based and brain stimulation therapies hold promise for providing a cellular environment conducive to promoting structural plasticity.

SUMMARY

A major question in experimental ischemia remains whether neuroprotection observed in young, healthy animals can be extrapolated to the human population, which is generally sick and aging. It is important to consider the impact of comorbidities such as hypertension and diabetes that are highly associated with cerebral ischemia. Various other risk factors have been observed, such as obesity, smoking, and excessive alcohol consumption. In addition to these important modifiable risk factors, it is important to consider age and gender when developing models to test new treatments for cerebral ischemia. Animal models require consideration of various modifiable (hypertension, diabetes, obesity) and nonmodifiable (age, gender, genetics) risk factors, which all interact with each other to add to the complexities of ischemia. Consideration of comorbidities, combined with ongoing research focused on complex intercellular interactions in the injured brain, hold promise for improved preclinical modeling and identification of protective strategies that may ultimately translate to patients.

The narrow focus on brain protection has led to missed research opportunities. It is likely that long-term functional recovery can be enhanced by combining potential protective strategies with therapies designed to allow recovery of function in both injured and uninjured brain regions. Functional plasticity is a mechanism by which neuronal networks that remain following an ischemic brain injury undergo structural and synaptic modifications to restore lost function.[50] Recent advances in basic neuroscience research following acquired brain injury has been complemented by advances in functional MRI that show changes in functional connectivity that are associated with motor impairments and spontaneous recovery.[51] These changes occur within affected and nonaffected territories and reflect a potential target for improving functional recovery in patients with stroke. Transmagnetic stimulation of affected and nonaffected hemispheres has been shown to be beneficial for recovery of motor function in patients with stroke and was associated with functional connectivity of both hemispheres.[52] Other stimulation methods, such as vagus nerve and deep brain stimulation, are also being investigated for their ability to provide functional benefit.[53] The mechanisms by which these methods work to improve functional recovery continue to be elucidated and include altered neuronal excitability, enhanced synaptic plasticity, and the release of neurotrophins and other modulatory factors.

REFERENCES

1. Arundine M, Tymianski M. Molecular mechanisms of glutamate-dependent neuro-degeneration in ischemia and traumatic brain injury. Cell Mol Life Sci 2004;61: 657–68.
2. López-Valdés HE, Clarkson AN, Ao Y, et al. Memantine enhances recovery from stroke. Stroke 2014;45:2093–100.
3. Cook DJ, Teves L, Tymianski M. Treatment of stroke with a PSD-95 inhibitor in the gyrencephalic primate brain. Nature 2012;483:213–7.
4. Simon R, Xiong Z. Acidotoxicity in brain ischaemia. Biochem Soc Trans 2006;34: 1356–61.
5. Pignataro G, Simon RP, Xiong ZG. Prolonged activation of ASIC1a and the time window for neuroprotection in cerebral ischaemia. Brain 2007;130:151–8.
6. Chan PH. Role of oxidants in ischemic brain damage. Stroke 1996;27:1124–9.
7. Traystman RJ, Kirsch JR, Koehler RC. Oxygen radical mechanisms of brain injury following ischemia and reperfusion. J Appl Physiol (1985) 1991;71:1185–95.
8. Diener HC, Lees KR, Lyden P, et al. NXY-059 for treatment of acute stroke: pooled analysis of the SAINT I and II Trials. Stroke 2008;39:1751–8.
9. Tobin MK, Bonds JA, Minshall RD, et al. Neurogenesis and inflammation after ischemic stroke: what is known and where we go from here. J Cereb Blood Flow Metab 2014;34:1573–84.
10. Iadecola C, Mihaela A. Cerebral ischemia and inflammation. Curr Opin Neurol 2001;14:89–94.
11. Elmore S. Apoptosis: a review of programmed cell death. Toxicol Pathol 2007;35: 495–516.
12. Moskowitz MA, Lo EH. Neurogenesis and apoptotic cell death. Stroke 2003;34: 324–6.
13. Herson PS, Traystman RJ. Animal models of stroke: translational potential at present and in 2050. Future Neurol 2014;9:541–51.
14. Traystman RJ, Herson PS. Misleading results: translational challenges. Science 2014;343:369–70.
15. Saver JL, Goyal M, Bonafe A, et al, SWIFT PRIME Investigators. Stent-retriever thrombectomy after intravenous t-PA vs. t-PA alone in stroke. N Engl J Med 2015;372(24):2285–95.
16. Berkhemer OA, van Zwam WH, Dippel DW. Stent-retriever thrombectomy for stroke. N Engl J Med 2015;373(11):1076–7.
17. Tissue plasminogen activator for acute ischemic stroke. The National Institute of Neurological Disorders and Stroke rt-PA Stroke Study Group. N Engl J Med 1995; 333(24):1581–7.
18. del Zoppo GJ, Mabuchi T. Cerebral microvessel responses to focal ischemia. J Cereb Blood Flow Metab 2003;23(8):879–94.
19. Simard M, Arcuino G, Takano T, et al. Signaling at the gliovascular interface. J Neurosci 2003;23(27):9254–62.
20. Nedergaard M, Ransom B, Goldman SA. New roles for astrocytes: redefining the functional architecture of the brain. Trends Neurosci 2003;26(10):523–30.
21. Thored P, Wood J, Arvidsson A, et al. Long-term neuroblast migration along blood vessels in an area with transient angiogenesis and increased vascularization after stroke. Stroke 2007;38(11):3032–9.

22. Ohab JJ, Fleming S, Blesch A, et al. A neurovascular niche for neurogenesis after stroke. J Neurosci 2006;26(50):13007–16.

23. Fernandez-Klett F, Potas JR, Hilpert D, et al. Early loss of pericytes and perivascular stromal cell-induced scar formation after stroke. J Cereb Blood Flow Metab 2013;33(3):428–39.

24. Fernandez-Klett F, Priller J. The fibrotic scar in neurological disorders. Brain Pathol 2014;24(4):404–13.

25. Pham LD, Hayakawa K, Seo JH, et al. Crosstalk between oligodendrocytes and cerebral endothelium contributes to vascular remodeling after white matter injury. Glia 2012;60(6):875–81.

26. Redecker C, Wang W, Fritschy JM, et al. Widespread and long-lasting alterations in GABA(A)-receptor subtypes after focal cortical infarcts in rats: mediation by NMDA-dependent processes. J Cereb Blood Flow Metab 2002;22(12):1463–75.

27. Clarkson AN, Huang BS, Macisaac SE, et al. Reducing excessive GABA-mediated tonic inhibition promotes functional recovery after stroke. Nature 2010;468(7321):305–9.

28. Lake EM, Chaudhuri J, Thomason L, et al. The effects of delayed reduction of tonic inhibition on ischemic lesion and sensorimotor function. J Cereb Blood Flow Metab 2015;35(10):1601–9.

29. Witgen BM, Lifshitz J, Smith ML, et al. Regional hippocampal alteration associated with cognitive deficit following experimental brain injury: a systems, network and cellular evaluation. Neuroscience 2005;133(1):1–15.

30. Kharlamov EA, Lepsveridze E, Meparishvili M, et al. Alterations of GABA(A) and glutamate receptor subunits and heat shock protein in rat hippocampus following traumatic brain injury and in posttraumatic epilepsy. Epilepsy Res 2011;95(1–2):20–34.

31. Mtchedlishvili Z, Lepsveridze E, Xu H, et al. Increase of GABAA receptor-mediated tonic inhibition in dentate granule cells after traumatic brain injury. Neurobiol Dis 2010;38(3):464–75.

32. Rehme AK, Eickhoff SB, Wang LE, et al. Dynamic causal modeling of cortical activity from the acute to the chronic stage after stroke. Neuroimage 2011;55(3):1147–58.

33. Clarkson AN, Overman JJ, Zhong S, et al. AMPA receptor-induced local brain-derived neurotrophic factor signaling mediates motor recovery after stroke. J Neurosci 2011;31(10):3766–75.

34. Dhawan J, Benveniste H, Luo Z, et al. A new look at glutamate and ischemia: NMDA agonist improves long-term functional outcome in a rat model of stroke. Future Neurol 2011;6:823–34.

35. Reeves TM, Lyeth BG, Povlishock JT. Long-term potentiation deficits and excitability changes following traumatic brain injury. Exp Brain Res 1995;106(2):248–56.

36. Li W, Huang R, Shetty RA, et al. Transient focal cerebral ischemia induces long-term cognitive function deficit in an experimental ischemic stroke model. Neurobiol Dis 2013;59:18–25.

37. Kiprianova I, Sandkühler J, Schwab S, et al. Brain-derived neurotrophic factor improves long-term potentiation and cognitive functions after transient forebrain ischemia in the rat. Exp Neurol 1999;159(2):511–9.

38. Sanders MJ, Sick TJ, Perez-Pinzon MA, et al. Chronic failure in the maintenance of long-term potentiation following fluid percussion injury in the rat. Brain Res 2000;861(1):69–76.

39. Orfila JE, Shimizu K, Garske AK, et al. Increasing small conductance Ca2+-activated potassium channel activity reverses ischemia-induced impairment of long-term potentiation. Eur J Neurosci 2014;40(8):3179–88.

40. Yaka R, Biegon A, Grigoriadis N, et al. D-cycloserine improves functional recovery and reinstates long-term potentiation (LTP) in a mouse model of closed head injury. FASEB J 2007;21(9):2033–41.

41. Nagy D, Kocsis K, Fuzik J, et al. Kainate postconditioning restores LTP in ischemic hippocampal CA1: onset-dependent second pathophysiological stress. Neuropharmacology 2011;61(5–6):1026–32.

42. Doyle KP, Quach LN, Solé M, et al. B-lymphocyte-mediated delayed cognitive impairment following stroke. J Neurosci 2015;35(5):2133–45.

43. Jones TA, Liput DJ, Maresh EL, et al. Use-dependent dendritic regrowth is limited after unilateral controlled cortical impact to the forelimb sensorimotor cortex. J Neurotrauma 2012;29(7):1455–68.

44. Hinman JD. The back and forth of axonal injury and repair after stroke. Curr Opin Neurol 2014;27:615–23.

45. Johnson VE, Stewart W, Smith DH. Axonal pathology in traumatic brain injury. Exp Neurol 2013;246:35–43.

46. Schwab ME, Strittmatter SM. Nogo limits neural plasticity and recovery from injury. Curr Opin Neurobiol 2014;27:53–60.

47. Armstrong RC, Mierzwa AJ, Marion CM, et al. White matter involvement after TBI: Clues to axon and myelin repair capacity. Exp Neurol 2015;275(Pt 3):328–33.

48. Prince DA, Parada I, Scalise K, et al. Epilepsy following cortical injury: cellular and molecular mechanisms as targets for potential prophylaxis. Epilepsia 2009;50(Suppl 2):30–40.

49. Wilson SM, Xiong W, Wang Y, et al. Prevention of posttraumatic axon sprouting by blocking collapsin response mediator protein 2-mediated neurite outgrowth and tubulin polymerization. Neuroscience 2012;210:451–66.

50. Caleo M. Rehabilitation and plasticity following stroke: Insights from rodent models. Neuroscience 2015;311:180–94.

51. Thiel A, Vahdat S. Structural and resting-state brain connectivity of motor networks after stroke. Stroke 2015;46(1):296–301.

52. Grefkes C, Fink GR. Disruption of motor network connectivity post-stroke and its noninvasive neuromodulation. Curr Opin Neurol 2012;25(6):670–5.

53. Cai PY, Bodhit A, Derequito R, et al. Vagus nerve stimulation in ischemic stroke: old wine in a new bottle. Front Neurol 2014;5:107.

Cerebral Blood Flow Autoregulation and Dysautoregulation

William M. Armstead, PhD[a,b],*

KEYWORDS

- Cerebral autoregulation • Traumatic brain injury • Anesthetics • Cerebral blood flow

KEY POINTS

- Cerebral autoregulation can be assessed by static and dynamic techniques.
- Inhaled anesthetics generally depress autoregulation, except for sevoflurane. Total intravenous anesthesia (TIVA) is most widely used in neurocritical care of traumatic brain injury patients.
- Autoregulation is impaired after traumatic brain injury. Cerebral perfusion pressure–directed therapy is often used to improve outcome.
- Identification of optimal cerebral perfusion pressure in health and disease is an active area of investigation.
- Role of age and sex in cerebral autoregulation in physiology and pathology of traumatic brain injury is an emerging area of clinical significance.

INTRODUCTION

Cerebral autoregulation is a homeostatic process that regulates and maintains cerebral blood flow (CBF) constant across a range of blood pressures. The original concept was proposed by Lassen[1] as a triphasic curve consisting of the lower limit, the plateau, and the upper limit. In healthy adults, the limits are between 50 and 150 mm Hg cerebral perfusion pressure (CPP) or 60 and 160 mm Hg mean arterial pressure (MAP), where CPP = MAP – intracranial pressure (ICP). This homeostatic mechanism ensures that as MAP or CPP increases, resistance increases (vasoconstriction) in the small cerebral arteries. Conversely, this process maintains constant CBF by decreasing cerebrovascular resistance or vasodilation when MAP or CPP decreases. However, given that the

This work is published in collaboration with the Society for Neuroscience in Anesthesiology and Critical Care.

Sources of Financial Support: NIH RO1 NS090998.

[a] Department of Anesthesiology and Critical Care, University of Pennsylvania, 3620 Hamilton Walk, JM3, Philadelphia, PA I9l04, USA; [b] Department of Pharmacology, University of Pennsylvania, Philadelphia, PA I9l04, USA

* Department of Anesthesiology and Critical Care, University of Pennsylvania, 3620 Hamilton Walk, JM3, Philadelphia, PA I9l04.

E-mail address: armsteaw@uphs.upenn.edu

lower limit of cerebral autoregulation frequently influences clinical management, it should be noted that this value has been challenged as being too low.[2]

MECHANISM OF AUTOREGULATION

At least 4 mechanisms are proposed for autoregulation:

1. Myogenic
2. Neurogenic
3. Metabolic
4. Endothelial

The myogenic component concerns the ability of the vascular smooth muscle to constrict or dilate in response to changes in transmural pressure. The neurogenic mechanism occurs through an extensive nerve supply to cerebral vessels. For example, activation of α-adrenoceptors shifts the limits of autoregulation to higher pressures, and the cerebral vessels respond to this with vasoconstriction. The metabolic mechanism probably chiefly contributes to autoregulation in the microvasculature, in which changes in the microenvironment such as for partial pressure of carbon dioxide (P_{CO_2}) and pH will lead to vasodilation. Additionally, endothelial factors, such as nitric oxide, may also contribute to autoregulation. Autoregulation is important in the match of CBF to metabolic demand.

METHODS USED FOR ASSESSMENT OF AUTOREGULATION

Assessment of cerebral autoregulation:

1. Static
2. Dynamic

In the static method, only steady-state relationships between CBF and MAP are considered without taking into account the time course of changes in these 2 parameters. Determination of a steady-state relationship can be accomplished through administration of drugs, which change MAP but have no effect on metabolism, yielding 2 values of CBF and their difference in relation to the MAP change being indicative of autoregulation.[3] Degree of intactness of autoregulation can be quantified via calculation of the autoregulatory index (ARI), where ARI = % ΔCVR/% ΔCPP and CVR is cerebrovascular resistance. A value of 0 means absent autoregulation, whereas a value of 1 denotes perfect autoregulation.

In the dynamic method, assessment is based on determination of dynamic changes of CBF in response to dynamic changes in MAP. The idea is that after a change in MAP, CBF will first react to such a change and then will return to its original value within a finite amount of time. The faster the return, the better is the degree of autoregulation.

Several approaches are used to assess dynamic cerebral autoregulation via CBF recovery time. In one, which uses the thigh cuff method introduced by Aaslid and colleagues,[4] regulation is defined by the slope of CVR recovery, where CVR = CPP/CBF; the steeper the slope of CVR, the better the autoregulation.[5] In a second approach, the CBF recovery after the cuff release is quantified as an autoregulation index, calculated as a second-order differential equation relating changes in MAP and CBF.[5] A third method considers study of CBF response to slow oscillations in MAP induced by paced breathing, head up tilting, or periodic thigh cuff inflation.[6–8] Transfer function analysis is then performed using beat-to-beat MAP measurements as input and CBF as output.[9] Here, time delay of phase difference between MAP and CBF as a function of frequency can be used to determine the degree of intactness of autoregulation. This approach is an important alternative, as the thigh cuff inflation/ deflation technique presents problems in the traumatic brain injury (TBI) patient, in which manipulation of blood pressure (secondary to thigh cuff deflation) would occur at a time when the injured brain may be least able to tolerate it. The mean index (Mx) is one such measure of dynamic autoregulation. Mx is a Pearson correlation coefficient between CPP and flow velocity measured by transcranial Doppler (TCD) and can indicate autoregulation if the magnitude of CPP fluctuations is large enough to activate an autoregulatory response (>5 mm Hg). Positive values for Mx indicate impaired autoregulation, whereas zero or negative values indicate intact autoregulation. This index correlates significantly with the thigh cuff test.[10] Another similar measure of dynamic autoregulation is pressure reactivity index (PRx), defined as the correlation coefficient between slow waves in ICP and arterial blood pressure. The PRx is negative for intact autoregulation, and is positive for impaired autoregulation.[11]

CBF in all of these methods is typically determined noninvasively via TCD. TCD, however, has the disadvantage of only being able to measure CBF velocity in larger vessels such as the middle cerebral artery. Recently, it was shown that use of near infrared spectroscopy will allow for measurement of CBF in small cerebral arterioles, yielding a quantitative determination of dynamic autoregulation in the cerebral microvasculature of the human.[12] A related technique, diffuse correlation spectroscopy (DCS), is found to obtain regional CBF noninvasively in patients under a variety of conditions, including TBI, orthostatic stress in acute ischemia, and postoperatively in neonatal cardiac surgery.[13,14] Calibration studies of absolute CBF values obtained via DCS have been recently completed,[13] making the future outlook for use of DCS as a quantitative method for determination of autoregulation in the microvasculature more attractive.

CLINICAL APPLICATIONS OF AUTOREGULATION TESTING

A benefit to autoregulation testing relates to management of patients with TBI in the neuro intensive care unit, where sophisticated equipment for determination of dynamic indices of autoregulation is available. However, there are tangible benefits for management of neurologic patients in the operating room through use of simpler determination of static autoregulation. These benefits include:

1. Perioperative management of patients undergoing carotid endarterectomy.
2. Blood pressure management in uncontrolled hypertensive patients.
3. Blood pressure support in patients with carotid artery stenosis.

EFFECT OF ANESTHETICS ON AUTOREGULATION

In general, inhaled anesthetics such as isoflurane and desflurane have a depressive effect on autoregulation.[15,16] One exception is sevoflurane.[17,18] In contrast, propofol preserves autoregulation[15] and is viewed as the anesthetic of choice in patients with high ICP,[19] although one study observed that high-dose propofol impaired autoregulation in head-injured patients.[20] Others have observed that the combination of remifentanil with propofol results in preservation of cerebral autoregulation.[19,21] Current neurocritical care of TBI is largely based on TIVA, or total intravenous anesthesia, often involving fentanyl.

PATHOPHYSIOLOGY OF CEREBRAL AUTOREGULATION IN TRAUMATIC BRAIN INJURY: OUTCOME AND TREATMENT

Multiple studies found that cerebral autoregulation is absent or impaired in significant numbers of patients after TBI, even when values of CPP and CBF were normal.[22]

Cerebral hypoperfusion that occurs after TBI may result from the decreased metabolic demand caused by coma (with appropriately matched CBF) or may alternatively indicate ischemia. Under normal physiologic conditions, decreases in CPP result in vasodilation and increased cerebral blood volume, leading to increases in ICP in the context of abnormal intracranial compliance. When autoregulation is impaired, decreases in CPP result in decreases in CBF; in moderate/severe TBI such decreases in CBF may reach ischemic levels, further exacerbating secondary injury. Several retrospective studies found that impaired cerebral autoregulation is associated with worsened outcome (Glasgow Outcome Scale).[23–25] Cerebral autoregulation is also found to correlate with brain biochemistry via microdialysis in patients with severe head injury.[26] Critical autoregulatory thresholds for survival and improvement of outcome may vary as a function of age and sex.[23,24]

One treatment modality has been derived with specific consideration of cerebral autoregulation. This treatment relies on administration of vasoactive agents that increase MAP to normalize reduced CPP after TBI and the identification of the optimal CPP that maximizes the intactness of autoregulation. Those patients with CPP values less than those considered optimal are found to be at increased risk of fatal outcome, whereas too high a value of CPP is equally thought to be bad, as it is often associated with increased rate of severe disability.[27] Optimal CPP, as determined by PRx in TBI patients, is found to be narrowed from a normal range and varies from 65 to 95 mm Hg in 300 patients (with total mean value of 75 mm Hg).[25] Similar results were obtained using other technologies to determine degree of intactness of autoregulation, such as direct brain tissue monitoring and near-infrared spectroscopy.[28,29] Although such results would seem to indicate that randomized, prospective trials involving CPP (optimal) are needed, to date no such trial has been conducted.

Given that dysregulation seems to correlate quite well with clinical outcome, one could hypothesize that autoregulation should be therapeutically manipulated to improve recovery.[25] Although this hypothesis has never been formally prospectively tested, a few observations have hinted that there may be support for it. For example, Mx value was observed to improve with induction of moderate hypocapnia in TBI patients,[30] but autoregulation was observed to be impaired when hyperventilation was more intense, resulting in a robust decrease in Pco_2.[31] Additionally, indomethacin is found to decrease ICP and cerebral blood flow velocity, thereby increasing CPP, resulting in improved dynamic cerebral autoregulation in still other TBI patients.[32] Other clinical work found that other therapeutic maneuvers equally change cerebral autoregulation status, such as hypothermia, use of mannitol or hypertonic saline, barbiturates, and various anesthetics.[25,33]

ROLE OF AGE AND SEX IN CEREBRAL AUTOREGULATION, DYSAUTOREGULATION, AND ITS TREATMENT AFTER TRAUMATIC BRAIN INJURY
Clinical Studies

Only a few studies of cerebral autoregulation in healthy children have been published, and clinicians often assume that there are no age- or sex-related differences in this process.[34] In one study, no age-related differences in autoregulatory capacity were found in healthy children anesthetized with low-dose sevoflurane.[35] However, counter to the assumption that the lower limit of cerebral autoregulation in younger children is lower, these investigators also concluded that there were no age-related differences in lower limit, as that value in children age 6 months to 2 years receiving sevoflurane was observed to be 60 ± 9 mm Hg.[36] Alternatively, there may be age-related differences in the latency of cerebral autoregulation. For example, compared with adults, adolescents may have a delayed return of CBF in response to brief hypotension.[37] Additionally, there may be sex differences in pediatric cerebral autoregulation.[34,37,38] Despite these clinical studies, the mechanisms of cerebral autoregulation in healthy children are not understood completely.[39]

Several conditions can disrupt autoregulation in children, including hypoxic conditions, prematurity, congenital heart defects, intracranial hemorrhage, and TBI. Risk of dysautoregulation in preterm infants increases with severity of illness.[40] TBI is the leading cause of injury-related death in children and young adults[41] and thus will be the focus here; boys older than a larger age range and all children younger than 4 years have particularly poor outcomes.[42–44] Cerebral autoregulation is often impaired after TBI,[23] and with concomitant high ICP, leads to poor outcome. In children with impaired autoregulation, lower blood pressure may result in diminished CPP and CBF. Decrease in MAP causes cerebral vasodilation, increase in cerebral blood volume, and, thus, an increase in ICP. In the context of plateau wave generation, increased ICP will further decrease CPP, leading to more cerebral vasodilation, resulting in a vicious cycle, the vasodilation cascade.[45] Critical closing pressure has been defined as an arterial pressure threshold below which arterial vessels collapse. Dewey and colleagues[46] identified in their model of CBF regulation a relationship between critical closing pressure and cerebral autoregulation indicating that cerebral autoregulation mediated increases in vasomotor tone as predicted by critical closing pressure. This concept has recently been revisited,[47] but although the difference in CPP-ICP significantly correlated with cerebral autoregulation, it lacked the power to predict outcome after head injury. Nonetheless, by stabilizing CPP at higher levels, ICP might be better controlled without cerebral ischemia.[45] Because augmenting MAP in the hyperemic brain could theoretically result in worse edema or cerebral hemorrhage,[44,48] it is uncertain if empirically increasing MAP to prevent cerebral ischemia in the presence of impaired cerebral autoregulation and cerebral hyperemia may potentially be harmful.[35] Moreover, a fixed blood pressure does not improve nor allow for cerebrovascular adaptation to changing CBF metabolism requirements. Because cerebroautoregulation changes with time after TBI, it is suggested that blood pressure management strategies can neither be empiric nor fixed but should be based on the status of cerebral autoregulation. In this context, vasodilator mechanisms of cerebral autoregulation may be important in improving outcome after TBI. The functional significance of this idea rests on several clinical observations. For example, cognitive outcome and risk of death are correlated with degree of cerebral autoregulatory impairment in children after TBI.[23,34] Additionally, severity of impaired cerebral autoregulation may be greater in those children who are victims of intentional or abusive TBI compared with children with noninflicted TBI.[49]

Current 2012 Pediatric Guidelines recommend maintaining CPP greater than 40 mm Hg, noting that an age-related continuum for the optimal CPP is between 40 and 65 mm Hg.[50,51] Despite these current therapeutic targets, guidelines are lacking regarding how this should be achieved. For example, maintaining CPP within these levels is often managed by use of vasopressors to increase CPP and optimize CBF. However, vasoactive agents clinically used to elevate MAP after TBI, such as phenylephrine (Phe), dopamine (DA), epinephrine and norepinephrine (NE)[52,53] have not sufficiently been compared regarding effect on CPP, CBF, autoregulation, and survival after TBI, and clinically, current vasopressor use is variable. A recent single-center study found that: (1) Phe was used nearly twice as often as any other vasopressor; (2) young children tended to receive DA and epinephrine, whereas older children received Phe and NE; (3) blood pressure was managed with 1 drug during the first 3 hours; and (4) NE was associated with a higher CPP and lower ICP 3 hours

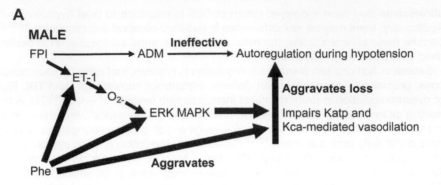

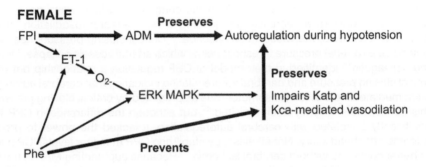

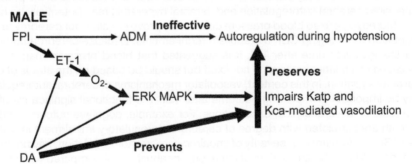

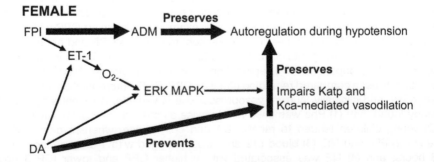

afterinjury.[54] Nonetheless, CPP-directed therapy has remained somewhat controversial because it has been observed to have no effect or worsen outcome.[55] Further complicating this issue is that most pressors are neurotransmitters, which normally have no effect on the brain when given peripherally. However, with a disrupted blood–brain barrier, these drugs may cross the blood–brain barrier and impact cerebral metabolism. Additionally, CPP has been considered a poor surrogate for CBF,[56] as regional or local CBF may be markedly reduced even if CPP is normal.[55] Therefore, there remains an unmet need to have available appropriate noninvasive bedside monitors, which may enable identification of the optimal vasopressor and whether pressor choice determines outcome as a function of age and sex in pediatric TBI.

Basic Science Studies

Although several rodent models of TBI have been described,[57] all have the disadvantage of not permitting repeated measurements of systemic physiologic variables and regional CBF because of the small size of the subjects. Additionally, rodents have a lissencephalic brain containing more grey than white matter. In contrast, pigs have a gyrencephalic brain similar to humans that contains substantial white matter, which is more sensitive to ischemic damage than grey matter.

Because ethical considerations constrain mechanistic studies in children with TBI, we have used an established porcine model of fluid percussion injury (FPI) that mimics TBI to corroborate clinical observations regarding cerebral autoregulation after TBI.[58] Newborn and juvenile pigs may approximate the human neonate (6 months–2 years old) and child (8–10 years old).[59]

Armstead and Vavilala[60] have recently taken a bidirectional translational approach to investigate the role of vasoactive agent choice in outcome after TBI. Clinical observations were used to inform study design of procedures conducted in pigs after FPI, with intent to use information so derived to improve outcome in brain-injured children, for example, a bedside to bench and back again arrangement. Early studies focused on Phe, given that it is the predominant vasoactive agent administered clinically in young children after TBI.[54] It was surprising to observe that although Phe is protective of cerebral hemodynamics, particularly for autoregulation, in newborn female piglets, it aggravates cerebrovascular dysregulation in male newborn piglets after injury.[60] Because of this perplexing observation, we speculated that pressor choice may influence outcome. Indeed, another pressor, DA, produced equivalent cerebrohemodynamic protection in both male and female piglets after TBI.[61]

Mechanistic studies were designed to further explore these observations. First, it was found that autoregulation was impaired more in the male compared with the female piglet after FPI, resulting from a shift in the constrictor/dilator ratio, with greater brain concentration of the spasmogen, endothelin-1 (ET-1), and little to no increase in the endogenous vasodilator peptide, adrenomedullin (ADM), a K channel opener (**Fig. 1**).[62,63] Autoregulation was previously observed to be dependent on opening of K channels (particularly atp sensitive and calcium sensitive K channels [Katp and Kca]), whereas ET-1 is known to impair opening of K channels via release of activated

Fig. 1. Comparison of proposed mechanisms for Phe (*A*) and DA (*B*) in control of cerebral hemodynamics after FPI in newborn piglets. Arrow thickness in proportion to probability of action. O_{2-}, oxygen free radical. (*From* Armstead WM, Vavilala, MS. Age and sex differences in hemodynamics in a large animal model of brain trauma. In: Lo EH, Lok J, Ning M, et al. editors. Vascular mechanisms in CNS trauma. New York: Springer; 2014. p. 278; with permission.)

oxygen, which then activates (phosphorylates) the extracellular signal-regulated kinase (ERK) isoform of mitogen-activated protein kinase (MAPK), a distal signaling system important in control of cerebral hemodynamics (see **Fig. 1**).[58] Because more ERK MAPK is released after FPI in the male compared with the female,[62,64] there is correspondingly greater impairment of autoregulation and Katp and Kca channel agonist–mediated cerebrovasodilation after injury in males compared with females.[65] Although Phe blunted ET-1 and ERK MAPK upregulation in female piglets after FPI, there was an unanticipated and unwanted Phe-mediated aggravation of ET-1 and ERK MAPK upregulation in male piglets after injury (see **Fig. 1**).[60] The latter compounded the already greater release of ET-1 and ERK MAPK in males compared to females after FPI and seemed to contribute to the sex-dependent impairment of autoregulation.[60] The ET-1 antagonist, BQ-123, blocked elevation of cerebrospinal fluid ERK MAPK and the aggravation of such elevation by Phe after FPI.[60] Coadministered BQ-123 with Phe also prevented impairment of autoregulation after FPI, supportive of the intermediary role for ET-1 in sex-dependent Phe-mediated hemodynamic dysregulation. Alternatively, Phe also promoted the increase in brain levels of ADM in females but not males after FPI.[60] In the context of ADM being an opener of K channels, Phe can be viewed as upregulating an endogenous neuroprotectant in females after FPI. Therefore, Phe augments impairment of autoregulation in males after TBI because it promotes release of a spasmogen and subsequent block of Katp and Kca channels because it simultaneously limits release of an endogenous neuroprotectant (see **Fig. 1**).[66] In contrast, DA completely blocked ET-1 and ERK MAPK upregulation in both newborn male and female piglets after FPI (see **Fig. 1**),[61] which explains the equivalent protection of autoregulation after injury in male and female piglets.

A third vasoactive drug, NE, has recently been investigated. In the first study, we observed that NE preferentially protected cerebral autoregulation and prevented neuronal cell necrosis in hippocampal areas CA1 and CA3 in female newborn piglets after FPI.[67] However, NE had no protective effect on cerebral autoregulation and potentiated neuronal cell necrosis in male newborn pigs, despite achievement of a similar CPP.[67] In the second study, NE protected cerebral autoregulation and limited hippocampal neuronal cell necrosis after FPI in both male and female juvenile pigs,[68] indicating that the same vasoactive drug elicits both age- and sex-dependent differences in outcome under equivalent injury intensity conditions.

Mechanistically, 2 mediators were investigated in the above studies: ERK MAPK and interleukin (IL)-6. Clinically, IL-6 was observed to be increased in the cerebral spinal fluid of children after severe TBI.[69] The role of IL-6 in central nervous system pathology, however, is less well understood. For example, IL-6 may mediate motor coordination deficits after TBI,[70] appears associated with poor neurologic outcome after hemorrhagic stroke,[71] and may also be involved in regenerative and repair processes.[69] In our first study, NE increased ERK MAPK upregulation in young males, but blocked upregulation in young females after FPI.[67] In contrast, in the second study using older pigs, NE blunted ERK MAPK and IL-6 upregulation in both males and females after FPI.[68] Coadministration of the ERK MAPK antagonist, U0126, preserved autoregulation and maintained CBF after FPI in newborns, regardless of sex.[67] U0126 also prevented hippocampal neuronal cell necrosis associated with FPI.[64] IL-6 was similarly increased after FPI more in young males compared with females, and NE aggravated such upregulation in an ERK MAPK–dependent manner in males; the ERK MAPK antagonist, U0126, blocked IL-6 release.[64] In contrast, NE blocked IL-6 upregulation in young females after FPI.[64] These data indicate that ERK MAPK mediates elevation of IL-6, which is associated with greater hemodynamic impairment and hippocampal cell necrosis after FPI in young males.

In the context of the neurovascular unit, we hypothesize that CBF contributes to neuronal cell integrity; for example, normalization of CBF via administration of vasoactive drugs that elevate MAP should limit neuronal cell necrosis associated with TBI. Administration of NE to increase CPP after TBI in the piglet prevented impairment of cerebral autoregulation and neuronal cell necrosis in the CA1 and CA3 hippocampal areas of the brain.[67] Clinically, degree of impairment of autoregulation is significantly correlated with Glasgow Coma Scale.[23,34] Although cognition depends on more than the hippocampus, and formal cognitive testing was not performed in these studies, these data nonetheless support the idea that targeting CPP and thereby protecting cerebral autoregulatory capacity may improve cognitive outcome. The neurovascular unit concept and its potential role in cognition may be somewhat more universal than prior appreciated. For example, administration of tPA-S481A, a catalytically inactive tPA variant that competes with wild-type tPA for binding, cleavage, and activation of N-Methyl-D-aspartate receptors, after TBI in the pig, similarly improved cerebral hemodynamics and prevented impairment of cerebral autoregulation and CA1 and CA3 hippocampal cell necrosis.[72] We speculate that choice of pressor may influence cognitive outcome as a function of age and sex in the setting of TBI.

Vasoactive agents clearly have complex sex-dependent effects on CBF and may potentiate impairment (Phe), prevent impairment (DA), or have no effect (NE) on impairment of cerebral autoregulation in newborn boys after TBI. Additionally, when considering age, NE may have no effect in the newborn, or prevent impairment in the juvenile male after TBI. These studies strengthen the idea that choice of vasoactive agent is important in determining outcome after pediatric TBI as a function of sex and age. To our knowledge, similar studies have not been conducted in the adult.

SUMMARY

This article emphasized a bidirectional translational research approach in summarizing state-of-the-art aspects of cerebral autoregulation that range from the traditional to the newly emerging. Enhanced cross-talk between basic science and clinical disciplines will speed the transfer of knowledge, resulting in better therapeutic approaches toward improving outcome in patients with central nervous system conditions such as TBI, in which there is cerebral dysautoregulation.

REFERENCES

1. Lassen LA. Cerebral blood flow and oxygen consumption in man. Physiol Rev 1959;39:183–238.
2. Drummond JC. The lower limit of autoregulation: time to revise our thinking? Anesthesiology 1997;86:1431–3.
3. Lassen LA. Control of cerebral circulation in health and disease. Circ Res 1974; 34:749–60.
4. Aaslid R, Lindegaard KF, Sorteberg W, et al. Cerebral autoregulation dynamics in humans. Stroke 1989;20:45–52.
5. Tiecks FP, Lam AM, Aaslid R, et al. Comparison of static and dynamic cerebral autoregulation measurements. Stroke 1995;26:1014–9.
6. Reinhard M, Wehrle-Wieland E, Grabiak D, et al. Oscillatory cerebral hemodynamics-the macro vs microvascular level. J Neurol Sci 2006;250:103–9.

7. Diehl RR, Linden D, Lucke D, et al. Phase relationship between cerebral blood flow velocity and blood pressure. A clinical test of autoregulation. Stroke 1995; 26:1801–4.

8. Aaslid R, Blaha M, Sviri G, et al. Asymmetric dynamic cerebral autoregulatory response to cyclic stimuli. Stroke 2007;38:1465–9.

9. Zhang R, Zuckerman JH, Giller CA, et al. Transfer function analysis of dynamic cerebral autoregulation in humans. Am J Physiol 1998;274:H233–41.

10. Czosnyka M, Smielewski P, Piechnik S, et al. Cerebral autoregulation following head injury. J Neurosurg 1974;95:756–63.

11. Steiner LA, Coles JP, Czosnyka M, et al. Cerebrovascular pressure reactivity is related to global cerebral oxygen metabolism after head injury. J Neurol Neurosurg Psychiatry 2003;74:765–70.

12. Kainerstorfer JM, Sassaroli A, Tgavalekos KT, et al. Cerebral autoregulation in the microvasculature measured with near-infrared spectroscopy. J Cereb Blood Flow Metab 2015;35:959–66.

13. Durduran T, Yodh AG. Diffuse correlation spectroscopy for non invasive, microvascular cerebral blood flow measurement. Neuroimage 2014;85:51–63.

14. Kim MN, Durduran T, Frangos S, et al. Noninvasive measurement of cerebral blood flow and blood oxygenation using near infrared and diffuse correlation spectroscopies in critically brain injured adults. Neurocrit Care 2010;12:173–80.

15. Strebel S, Lam AM, Batta B, et al. Dynamic and static cerebral autoregulation during isoflurane, desflurane, and propofol anesthesia. Anesthesiology 1995;83: 66–76.

16. Dagal A, Lam AM. Cerebral autoregulation and anesthesia. Curr Opin Anaesthesiol 2009;22:547–52.

17. Summors AC, Gupta AK, Mat BF. Dynamic cerebral autoregulation during sevoflurane anesthesia: a comparison with isoflurane. Anesth Analg 1999;88:341–5.

18. Gupta S, Heath K, Matta BF. Effect of incremental does of sevoflurane on cerebral pressure autoregulation in humans. Br J Anaesth 1997;79:469–72.

19. Cole CD, Gottfried ON, Gupta DK, et al. Total intravenous anesthesia: advantages for intracranial surgery. Neurosurgery 2007;61(Suppl 2):369–77.

20. Grathwohl KW, Black IH, Spinella PC, et al. Total intravenous anesthesia including ketamine versus volatile gas anesthesia for combat-related operative traumatic brain injury. Anesthesiology 2008;109:44–53.

21. Engelhard K, Werner C, Mollenberg O, et al. Effects of remifentanil/propofol in comparison with isoflurane on dynamic cerebrovascular autoregulation in humans. Acta Anaesthesiol Scand 2001;45:9971–6.

22. Rangel-Castilla L, Gasco J, Nauta HJ, et al. Cerebral pressure autoregulation in traumatic brain injury. Neurosurg Focus 2008;25:1–8.

23. Freeman SS, Udomphorn Y, Armstead WM, et al. Young age as a risk factor for impaired cerebral autoregulation after moderate-severe pediatric brain injury. Anesthesiology 2008;108:588–95.

24. Sorrentino E, Diedler J, Kasprowicz M, et al. Critical thresholds for cerebrovascular reactivity after traumatic brain injury. Neurocrit Care 2012;16:258–66.

25. Czonsyka M, Miller C. Monitoring of cerebral autoregulation. Neurocrit Care 2014; 21:S95–102.

26. Chan TV, Ng SC, Lam JM, et al. Monitoring of autoregulation using intracerebral microdialysis in patients with severe head injury. Acta Neurochir Suppl 2005;95: 113–6.

27. Aries MJ, Czosnyka M, Budohoski KP, et al. Continuous determination of optimal cerebral perfusion pressure in traumatic brain injury. Crit Care Med 2012;40: 2456–63.

28. Jaeger M, Schuhmann MU, Soehle M, et al. Continuous assessment of cerebrovascular autoregulation after traumatic brain injury using brain tissue oxygen pressure reactivity. Crit Care Med 2006;34:1783–8.

29. Zweifel C, Castellani G, Czosnyka M, et al. Noninvasive monitoring of cerebrovascular reactivity with near infrared spectroscopy in head-injured patients. J Neurotrauma 2010;27:1951–8.

30. Haubrich C, Steiner L, Kim DJ, et al. How does moderate hypocapnia affect cerebral autoregulation in response to changes in perfusion pressure in TBI patients? Acta Neurochir Suppl 2012;114:153–6.

31. Newell DW, Weber JP, Watson R, et al. Effect of transient moderate hyperventilation on dynamic cerebral autoregulation after severe head injury. Neurosurgery 1996;39:35–43.

32. Puppo C, Lopez L, Caragna E, et al. Indomethacin and cerebral autoregulation in severe head injured patients: a transcranial Doppler study. Acta Neurochir (Wien) 2007;149:139–49.

33. Steiner LA, Johnston AJ, Chatfield DA, et al. The effects of large dose propofol on cerebrovascular pressure autoregulation in head injured patients. Anesth Analg 2003;97:572–6.

34. Udomphorn Y, Armstead WM, Vavilala MS. Cerebral blood flow and autoregulation after pediatric traumatic brain injury. Pediatr Neurol 2008;38:225–34.

35. Vavilala MS, Lee LA, Lee M, et al. Cerebral autoregulation in children during sevoflurane anesthesia. Br J Anaesth 2003;90:636–41.

36. Vavilala MS, Lee LA, Lam AM. The lower limit of cerebral autoregulation in children during sevoflurane anesthesia. J Neurosurg Anesthesiol 2003;15:307–12.

37. Vavilala MS, Newell DW, Junger E, et al. Dynamic cerebral autoregulation in healthy adolescents. Acta Anaesthesiol Scand 2002;46:393–7.

38. Vavilala MS, Kincaid MS, Muangman SL, et al. Gender differences in cerebral blood flow velocity and autoregulation between the anterior and posterior circulations in health children. Pediatr Res 2005;58:574–8.

39. Philip S, Udomphorn Y, Kirkham FJ, et al. Cerebrovascular pathophysiology in pediatric traumatic brain injury. J Trauma 2009;67:S128–34.

40. Williams M, Lee JK. Intraoperative blood pressure and cerebral perfusion: strategies to clarify hemodynamics. Paediatr Anaesth 2014;24:657–67.

41. Au AK, Carcillo JA, Clark RS, et al. Brain injuries and neurological system failure are the most common proximate causes of death in children admitted to a pediatric intensive care unit. Pediatr Crit Care Med 2011;12:566–71.

42. Langlois JA, Rutland-Brown W, Thomas KE. The incidence of traumatic brain injury among children in the United States: differences by race. J Head Trauma Rehabil 2005;20:229–38.

43. Newacheck PW, Inkelas M, Kim SE. Heath services use and health care expenditures for children with disabilities. Pediatrics 2004;114:79–85.

44. Mandera M, Larysz D, Wojtacha M. Changes in cerebral hemodynamics assess by transcranial Doppler ultrasonography in children after head injury. Childs Nerv Syst 2002;18:124–8.

45. Rosner MJ, Rosner SD, Johnson AH. Cerebral perfusion pressure: management protocol and clinical results. J Neurosurg 1995;83:949–62.

46. Dewey RC, Pieper HP, Hunt WE. Experimental cerebral hemodynamics. Vaso-motor tone, critical closing pressure, and vascular bed resistance. J Neurosurg 1974;41:597–606.

47. Czosnyka M, Smielewski P, Piechnik S, et al. Critical closing pressure in cerebro-vascular circulation. J Neurol Neurosurg Psychiatry 1999;66:606–11.

48. Aldrich EF, Eisenberg HM, Saydjari C, et al. Diffuse brain swelling in severely head-injured children: a report from the NIH Traumatic Coma Data Bank. J Neurosurg 1992;76:450–4.

49. Vavilala MS, Muangman S, Waitayawinu P, et al. Neurointensive care; impaired cerebral autoregulation in infants and young children early after inflicted trau-matic brain injury: a preliminary report. J Neurotrauma 2007;24:87–96.

50. Carter BG, Butt W, Taylor A. ICP and CPP: excellent predictors of long term outcome in severely brain injured children. Childs Nerv Syst 2008;24:245–51.

51. Kochanek PM, Carney N, Adelson PD, et al. Guidelines for the acute medical management of severe traumatic brain injury in infants, children, and adolescents-second edition. Pediatr Crit Care Med 2012;13(Suppl 1):S24–9.

52. Biestro A, Barrios E, Baraibar J, et al. Use of vasopressors to raise cerebral perfu-sion pressure in head injured patients. Acta Neurochir 1998;71:5–9.

53. Steiner LA, Johnston AJ, Czosnyka M, et al. Direct comparison of cerebrovascu-lar effects of norepinephrine and dopamine in head injured patients. Crit Care Med 2004;32:1049–54.

54. DiGennaro JL, Mack CD, Malkouti A, et al. Use and effect of vasopressors after pediatric brain injury. Dev Neurosci 2010;32:420–30.

55. Coles JP, Steiner A, Johnston AJ, et al. Does induced hypertension reduce cere-bral ischemia within traumatized human brain? Brain 2004;127:2479–90.

56. Cremer OL, van Dijk GW, van Wensen E, et al. Effect of intracranial pressure monitoring and targeted intensive care on functional outcome after severe head injury. Crit Care Med 2005;33(10):2207–13.

57. Robertson CL, Scafidi S, McKenna MC, et al. Mitochondrial mechanisms of cell death and neuroprotection in pediatric ischemic and traumatic brain injury. Exp Neurol 2009;218:371–80.

58. Armstead WM. Age dependent cerebral hemodynamic effects of traumatic brain injury in newborn and juvenile pigs. Microcirculation 2000;7:225–35.

59. Dobbing J. The later development of the brain and its vulnerability. In: Davis JA, Dobbing J, editors. Scientific foundations of pediatrics. London: Heineman Med-ical; 1981. p. 744–59.

60. Armstead WM, Kiessling JW, Kofke WA, et al. Impaired cerebral blood flow autor-egulation during post traumatic arterial hypotension after fluid percussion brain injury is prevented by phenylephrine in female but exacerbated in male piglets by ERK MAPK upregulation. Crit Care Med 2010;38:1868–74.

61. Armstead WM, Riley J, Vavilala MS. Dopamine prevents impairment of autoregu-lation after TBI in the newborn pig through inhibition of upregulation of ET-1 and ERK MAPK. Pediatr Crit Care Med 2013;14:e103–11.

62. Armstead WM, Kiessling JW, Bdeir K, et al. Adrenomedullin prevents sex depen-dent impairment of autoregulation during hypotension after piglet brain injury through inhibition of ERK MAPK upregulation. J Neurotrauma 2010;27:391–402.

63. Armstead WM, Vavilala MS. Adrenomedullin reduces gender dependent loss of hypotensive cerebrovasodilation after newborn brain injury through activation of ATP- dependent K channels. J Cereb Blood Flow Metab 2007;27:1702–9.

64. Armstead WM, Riley J, Vavilala MS. TBI sex dependently upregulates ET-1 to impair autoregulation which is aggravated by phenylephrine in males but is abrogated in females. J Neurotrauma 2012;29:1483–90.
65. Armstead WM, Vavilala MS. Age and sex differences in hemodynamics in a large animal model of brain trauma. In: Lo EH, Lok J, Ning M, et al, editors. Vascular mechanisms in CNS trauma. New York: Springer; 2014. p. 269–84.
66. Armstead WM, Kiessling JW, Riley J, et al. Phenylephrine infusion prevents impairment of ATP and Calcium sensitive K channel mediated cerebrovasodilation after brain injury in female but aggravates impairment in male piglets through modulation of ERK MAPK upregulation. J Neurotrauma 2011;28:105–11.
67. Armstead WM, Riley J, Vavilala MS. Preferential protection of cerebral autoregulation and reduction of hippocampal necrosis with norepinephrine after traumatic brain injury in female piglets. Pediatr Crit Care Med 2016;17(3):e130–7.
68. Armstead WM, Riley J, Vavilala MS. Norepinephrine protects cerebral autoregulation and reduces hippocampal necrosis after traumatic brain injury via block of ERK MAPK and IL-6 in juvenile pigs. J Neurotrauma 2016. [Epub ahead of print].
69. Bell MJ, Kochanek PM, Doughty LA, et al. Interleukin-6 and interleukin-10 in cerebrospinal fluid after severe traumatic brain injury in children. J Neurotrauma 1997;14:451–7.
70. Yang SH, Gangidine M, Pritts TA, et al. Interleuking 6 mediates neuroinflammation and motor coordination deficits after mild traumatic brain injury and brief hypoxia in mice. Shock 2013;40:471–5.
71. Oto J, Suzue A, Inui D, et al. Plasma proinflammatory and anti-inflammatory cytokine and catecholamine concentrations as predictors of neurological outcome in acute strokepatients. J Anesth 2008;22:207–12.
72. Armstead WM, Riley J, Yarovoi S, et al. tPA S-481A prevents neurotoxicity of endogenous tPA in traumatic brain injury. J Neurotrauma 2012;29:1794–802.

Anesthesia for Neurosurgery and Interventional Radiology: Dhanesh K. Gupta

Chronic Pain in Neurosurgery

Samuel Grodofsky, MD

KEYWORDS

- Chronic pain • Neurosurgery • Multimodal analgesia • Spinal surgery • Craniotomy
- Adjunctive analgesics • Opioid sparing

KEY POINTS

- Chronic pain is an experience resulting from a wide variety of derangements leading to the abnormal processing of pain.
- Providers caring for patients with chronic pain undergoing neurosurgery should use multi-modal analgesia strategies to improve perioperative comfort and reduce the risk of post-operative sedation.
- There are few studies that have investigated specific anesthetic approaches for patients with chronic pain, and this requires a personalized approach to perioperative care.

INTRODUCTION

Providing safe and effective care for patients with chronic pain undergoing neurosurgery requires a carefully planned anesthetic strategy. If the anesthesiologist errs on the side of safety through conservative analgesic dosing, patients may emerge with harrowing pain, accompanied by unstable hemodynamics that may compromise the delicate operation. On the other hand, if the provider overdoses patients with opioids and other sedatives, patients may be too impaired to comply with the postoperative neurologic examination, may sustain respiratory failure, or the adrenergic response may be blunted and, thus, compromise perfusion to the spinal cord or brain.

Chronic pain has an estimated prevalence of 30% of the general population or 100 million people in the United States.[1,2] A proliferation of opioid prescribing has paralleled this epidemic as there has been a 4-fold increase in opioid prescriptions from 1999 to 2010.[3] A 2011 to 2012 survey reported 6.9% of adults taking an opioid in the last 30 days.[4] The implications of chronic opioid therapy (COT) and the pathophysiology of complex pain processing are separate but have overlapping clinical challenges. Pain is first and foremost an experience and should be understood in a biopsychosocial context that is influenced by musculoskeletal, nervous system,

This work is published in collaboration with the Society for Neuroscience in Anesthesiology and Critical Care.
The author of this publication has no financial disclosures.
Department of Anesthesiology and Critical Care, Hospital of the University of Pennsylvania, 3400 Spruce Street 5th Floor Dulles, Philadelphia, PA 19104, USA
E-mail address: Samuel.grodofsky@uphs.upenn.edu

Anesthesiology Clin 34 (2016) 479–495
http://dx.doi.org/10.1016/j.anclin.2016.04.003   **anesthesiology.theclinics.com**

emotional, and environmental factors. Patients undergoing craniotomies for tumor staging or debulking may present with diffuse, severe cancer-related pain and patients with abnormal spine conditions may have associated chronic pain. Not surprisingly, there is a particularly high occurrence of chronic pain in patients undergoing lumbar and cervical surgery.[5]

There is an association with altered pain perception and patients living with chronic pain. Experimental human studies have measured lower pain thresholds with pressure,[6] cold,[7] and heat[8] stimuli in chronic pain conditions compared with healthy controls. Prediction models consistently document chronic or preexisting pain as an independent risk factor for poorly controlled postoperative pain.[9–11] Interestingly, Chapman and colleagues[12] demonstrated that patients with chronic pain on COT reported clinically significant greater pain levels up to 15 days postoperatively compared with patients with chronic pain off long-term opioids.

Investigations into models of pain continue to evolve as newer evidence elaborates on older theoretic mechanisms. A quantum leap in the field occurred in 1965, when Melzack and Wall[13] published the gate control theory of pain (**Fig. 1**A). Previously, pain was described as a unidirectional pathway that was first transduced by small nociceptive peripheral nerves (C- and A-delta fibers) that were transmitted to afferent pathways along the spinal thalamic tract in the spinal cord before synapsing on neurons in the brain where pain perception occurs. According to the gate control theory, pain is not an unimpeded pathway from periphery to brain but results from an interaction between ascending afferent input carried by small-diameter nerve fibers and descending, inhibitory input carried by larger-diameter nerve fibers. Specialized neurons in the brain and spinal cord create an elaborate neural network that exerts a tonic inhibitory effect on interneurons that synapse in the substantia gelatinosa of the dorsal horn. According to Melzack and Wall,[13] "the substantia gelatinosa acts as a gate control system that modulates the synaptic transmission of nerve impulses from peripheral fibers to central cells."[13]

Various pain pathologies may be explained by an abnormally opened gate, where chronic inflammation or other neurologic derangements may create an imbalance of excitatory and inhibitory signaling. A windup of the dorsal horn has been demonstrated in some neuropathic pain states, which may give rise to the clinical findings of hyperalgesia or allodynia, whereby patients show an exaggerated response to light noxious or non-noxious stimuli, respectively.[14]

Decades after the gate control theory was introduced, researchers, including Melzack, have shifted attention from the dorsal horn to the brain. A new model proposes the presence of a "pain neuromatrix", which describes sensory and cognitive processes imprinting a representation of the "body-self" through a widely distributed neural network (**Fig. 1**B).[15,16] In this view, chronic pain is the output of the pain neuromatrix whereby nociceptive and non-nociceptive input results in an amplified experience of hurting.[17] Supporting this theory, brain imaging studies have correlated structural and functional differences in the anterior cingulate, prefrontal cortex, insula, somatosensory cortex, and parahippocampal gyrus in patients with fibromyalgia and central sensitization.[18,19]

Anesthesiologists must understand that pharmacologic regimens, which predictably treat acute pain in most adults, may fail in certain patients with complex derangements of their pain processing network. This article reviews perioperative management principles for chronic pain and focuses on clinical evidence emerging from neurosurgical studies. It is hoped that the reader will enrich his or her understanding of chronic pain and identify a broader palette of strategies that may be used to personalize a multimodal plan for patients undergoing a spinal or cranial procedure.

Another caveat to care pertains to the prevention of chronic pain developing from an acute surgical stimulus. This subject has become a hot area of investigation in many surgical procedures, including cranial and spinal surgeries.[20] Even cranial surgery, which historically was considered a low-pain-burden procedure, has been associated with a high incidence of chronic postsurgical headache.[21] The present state of evidence has not identified the precise genetic, environmental, and intraoperative management factors to predict the transition from acute to chronic pain; but there is a belief that providing superlative acute pain care can lower the risk of developing this crippling long-term comorbidity.[22]

PATIENT EVALUATION

Surgery introduces a new acute pain source that will be superimposed on a chronic pain experience. Evaluating the surgical plan may help anesthesiologists better prepare for the preemptive treatment of the noxious stimuli. For spinal surgeries, a complex, open, multilevel fusion that requires a larger incision; more dissection and injury to soft tissue, fascia, paraspinal muscles, and ligaments; and manipulation of dura, nerve roots, intravertebral discs, zygapophyseal facet joints, and other bony structures creates a greater pain stimulus in the immediate recovery period than a fusion performed through minimally invasive techniques.[23–25] In addition, the harvesting of patients' iliac crest bone to serve as a graft for fusion has also been associated with a greater pain burden.[26] For craniotomies, infratentorial procedures are associated with greater pain reports compared with supratentorial approaches.[27] In posterior fossa surgery, a cranioplasty after a craniectomy has been documented to increase the risk of postoperative headache compared with craniotomy alone.[28,29]

Providers should also evaluate unique patient factors related to their pain experience. Neuropathic pain pathologies may be associated with preexisting nerve damage and providers should perform a targeted history and physical examination to investigate motor and sensory deficits. In addition, the vegetative effects of chronic pain may promote the avoidance of healthy habits and thereby lead to a decreased cardiopulmonary reserve.

Studies have identified prognostic indicators associated with poor long-term outcomes occurring with maladaptive behavioral pain behaviors, such as COT, particularly at high doses (>100 mg a day of morphine equivalents), low socioeconomic status and support, unemployment, poorly controlled depression, fear of movement (kinesophobia or fear avoidant behaviors), catastrophizing, anxiety, and substance use disorders.[30,31] The presence of these factors can help providers risk stratify care. Paying attention to patients' psychological infrastructure can also enhance communication, reduce preoperative anxiety, and better define postoperative pain expectations. It is unfortunately common to provide a negative connotation for chronic pain patients. Instead, it is more constructive to develop empathy to the fact that this person hurts every day. In this perspective, common maladaptive behavioral responses are symptoms of a chronic illness, rather than an easily modifiable character flaw.

Patients with chronic pain may have long medication lists that may include gabapentin, pregabalin, antidepressants, anticonvulsants, nonsteroidal antiinflammatory drugs (NSAIDs), muscle relaxants, and opioid medications. **Table 1** lists common adjunctive pain medications and some anesthetic implications.

Opioid management is covered in detail in a later section, but it is critical to identify patients with COT preoperatively. During the patient evaluation, providers should perform a chart review and discuss with patients the frequency of breakthrough opioid

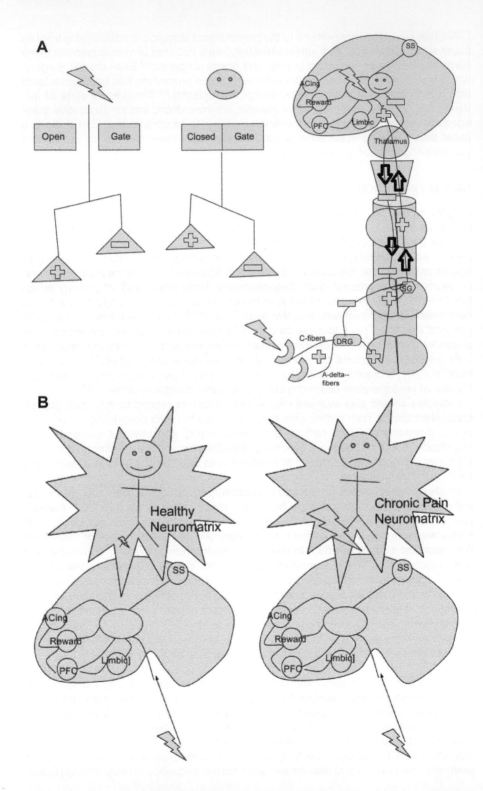

Table 1
Nonopioid medications used for the management of chronic pain and their anesthetic implications

Medication	Anesthetic Implications
Tricyclic antidepressants • Amitriptyline • Nortriptyline • Desipramine	• There are antimuscarinic, antihistaminergic, and anti-α_1-adrenergic effects. ○ Hypotension, constipation, sedation/delirium, dry mouth • It may reduce the seizure threshold. • EKG effect are as follows: QRS widening, QT prolongation • There is potentiation of sympathomimetic medications like ephedrine.
Selective serotonin norepinephrine reuptake inhibitors • Duloxetine • Venlafaxine • Milnacipran	• It may increase the risk of PONV. • It may increase blood pressure (mild effect in chronic use). • There is a slight bleeding tendency via decreased platelet binding affinity.
Selective serotonin reuptake inhibitors • Sertraline • Escitalopram	• There is a slight bleeding tendency via decreased platelet binding affinity. • It may increase the risk of PONV.
Calcium-channel blocking agents • Gabapentin • Pregabalin	• It may prevent an increase in intraocular pressure and blood pressure with direct laryngoscopy. • It may increase sedation.
Antiepileptic drugs • Oxcarbazepine • Carbamazepine • Topiramate	• There is increased resistance to nondepolarizing neuromuscular blockers (more frequent dosing). • There is a predisposition to hyponatremia. • May increase sedation • May induce hepatic microsomal enzymes leading to decreased plasma levels of various drugs, such as macrolide antibiotics, beta-blockers, calcium channel blockers, and amiodarone
Muscle relaxants • Cyclobenzaprine • Carisoprodol • Tizanidine • Baclofen • Methocarbamol	• There is increased sensitivity to neuromuscular blockers. • Cyclobenzaprine has antimuscarinic, antihistaminergic, and anti-α_1-adrenergic effects. • Methocarbamol may exacerbate myasthenia gravis symptoms. • It may increase sedation. • Tizanidine may further reduce blood pressure. • It may reduce minimum alveolar concentration.

Abbreviations: EKG, electrocardiogram; PONV, postoperative nausea and vomiting.

Fig. 1. (*A*) Model of the gate control theory of pain. Ascending nociceptive stimulus is transferred by C- and A-delta fibers to the dorsal root ganglion (DRG), substantia gelatinosa (SG), and then the spinal thalamic tract (and spinoreticular tract). When it reaches the brain, it is processed by the anterior cingulate (ACing), prefrontal cortex (PFC), somatosensory cortex (SS), limbic, reward centers, and other networks. Inhibitory signals carried by large-diameter fibers bind to the SG and prevent nociceptive signals from traveling centrally. The balance of inhibitory and excitatory input contributes to the transmission and experience of pain. (*B*) Model of the pain neuromatrix. A distributed neural network in the brain imprints a perception of the body-self. Pain results from a combination of sensory and cognitive input that projects the stimulus onto the neuromatrix.

dosing. This information can be used to calculate daily maintenance opioid requirements measured in oral morphine equivalents (milligrams per day). **Table 2** lists common conversion factors.

Spinal cord stimulators (SCS) and intrathecal drug delivery systems (IDDSs) are emerging technologies for a growing number of chronic pain conditions. SCS systems consist of electronic leads placed in the epidural space connected to an implantable pulse generator that is buried subcutaneously in the flank or buttocks. Patients are able to control the stimulator and should disable the device preoperatively if undergoing general anesthesia (GA). IDDSs consist of a flexible catheter placed in the intrathecal space that is attached to a medication reservoir pump that is implanted subcutaneously in the lower abdominal quadrant. Providers should continue the infusion perioperatively to maximize analgesic benefits. Neuraxial anesthesia is possible in the presence of these devices, but providers will need to investigate the spinal levels where the leads or catheters are placed. Spinal anesthesia should be uncomplicated assuming the intrathecal space is accessed at a lower intervertebral level than the device. If an epidural catheter is desired in the setting of SCS, coordination with a pain specialist may be warranted to avoid lead migration.

PHARMACOLOGIC TREATMENT OPTIONS

There are few studies in patients with chronic pain undergoing surgery. Because there are no best practice guidelines, anesthesiologists must optimize and individualize a care strategy using multiple modalities and mechanisms of action. Multimodal analgesia is the hallmark of perioperative care and is the preferred approach advocated by the American Society of Anesthesia (ASA) acute pain task force.[32] By combining multiple pharmacologic classes and procedural interventions during perioperative care, the total dose of any single agent may be lowered, thereby decreasing the side effect burden. In addition, this approach leverages the additive or synergistic effects when coadministrating different drug classes. In spinal surgery, several studies have demonstrated superior analgesia when using a multimodal analgesia strategy (**Table 3**) compared with opioid only controls.[33–35] This section summarizes the evidence of various medication classes and highlights neurosurgical trials to identify options for treatment strategies that can be personalized for patients with chronic pain.

Table 2
Equianalgesic opioid dosing

Drug	Equianalgesic Doses (mg)	
	Parenteral	Oral
Morphine	10	30
Hydromorphone	1.5	7.5
Oxycodone	NA	20
Fentanyl IV**	0.1	NA
Hydrocodone	NA	30
Methadone	10	10
Oxymorphone	1	10
Nalbuphine	10	10

Abbreviations: IV, intravenous; NA, not applicable .
** Fentanyl patch conversion: According to package insert, fentanyl transdermal 12 mcg/h = 45 mg/24 hours of oral morphine.

Table 3
Summary of adjunctive medications for establishing a multimodal analgesia strategy

Medication, Dosing Strategy	Issues
NSAIDs Ketorolac 30–60 mg IV[a] Celecoxib 200–400 mg PO[b]	• It is contraindicated in patients with compromised renal function. • It is used with caution in patients with a history of CAD or previous CVA. • Avoid it with concomitant antiplatelet, anticoagulation.
Acetaminophen 1000 mg IV[b], PO[a], and PR[b]	Make sure patients are not taking high doses the day of surgery.
Ketamine 0.5 mg/kg bolus IV, infusion 2–10 mcg/kg/min[b]	• Use caution with a history of psychosis or significant psychiatric illness. • It may cause nystagmus and increased intraocular pressure. • It may increase blood pressure and heart rate. • It may increase muscle tone or induce myoclonus.
Dexmedetomidine 1 mcg/kg IV bolus over 10 min, infusion at 0.2–1.0 mcg/kg/h[b]	• It may induce transient hypertension during bolus and then bradycardia, hypotension with infusion. • Use caution with concomitant use of vasodilators of negative chronotropic agents.
Gabapentin 600 mg PO[a] Pregabalin 150–300 mg PO[a]	• It may increase sedation after surgery. • It may increase risk of visual disturbance.
Lidocaine 1.5 mg/kg bolus IV, infusion at 2.0 mg/kg/h[b]	• Toxic doses may predispose to seizures and cardiovascular collapse. • It lowers the seizure threshold (pertinent in cortical surgery).

Dexamethasone IV 8.0 mg or 0.1 mg/kg[b]	• It increases the risk for the following:
	  ○ Infection
	  ○ GI bleed
	  ○ Blood glucose elevation
	  ○ Thromboembolism
	  ○ Delirium, psychosis, depression, and anxiety

Methadone (IV and PO) 0.2 mg/kg[ba]	• Avoid it if QTc is >475 ms. • There is an unpredictable pharmacokinetic half-life that may lead to delayed respiratory depression.

Abbreviations: CAD, coronary artery disease; CVA, cerebral vascular disease; GI, gastrointestinal; PR, per rectum.

[a] Dosing before induction.

[b] Dosing intraoperatively; use caution to complete infusions to facilitate emergence in a timely manner.

NONSTEROIDAL ANTIINFLAMMATORY DRUGS AND ACETAMINOPHEN

Acetaminophen may be administered orally, rectally, and intravenously (IV). A Cochrane review reported that a single IV dose is associated with a 30% decrease in opioid use and that approximately 37% of patients report adequate analgesia postoperatively for approximately 4 hours.[36] Although some hospitals may not carry IV formulations because of the expense, providers could consider rectal or oral administration.

NSAIDs, both traditional and selective cyclooxygenase-2 inhibitors, also demonstrate clinical efficacy.[35,37] A small prospective double-blind controlled study in patients undergoing lumbar decompressive surgery demonstrated that intraoperative ketorolac was associated with significant opioid sparing and lower pain scores throughout the entire postoperative recovery period.[38]

NSAIDs are often avoided in cranial and spinal surgery because of the known risk of bleeding, cardiovascular events, and impaired hardware to bone fusion. Much of this risk is exposed in animal models and medical datasets that reflect chronic use.[39,40] In addition, the bleeding association is typically associated with the coadministration with antiplatelet or anticoagulation therapy.[41] Most studies in surgical settings, however, support a favorable perioperative safety profile when selectively used in well-resuscitated patients. Richardson and colleagues[42] and Magni and colleagues[43] published retrospective reviews of pediatric and adult neurosurgeries, respectively, demonstrating that ketorolac was not associated with an increased risk for bleeding. Regarding the inhibition of osteoblast and bone metabolism, NSAIDs at low doses and a short time exposure (<14 days) have shown a strong safety profile when used with spinal fusion.[35]

Ketamine

Ketamine has been investigated as a therapeutic adjuvant for the chronic management of neuropathic pain,[44] complex regional pain syndrome,[45–47] and as a strategy to reduce daily opioid requirements.[48] Although infusions may show promise in the outpatient clinical setting, there is a lack of robust controlled trials and the literature is mixed.[46,49] In the perioperative setting, on the other hand, there is consistent evidence of benefit.[50–53] It has been shown particularly useful in patients with chronic pain[54] or opioid abuse disorders.[55]

Loftus and colleagues[54] investigated the use of ketamine in patients with COT undergoing lumbar fusion. An opioid-sparing effect and improved pain reports were found to be clinically and statistically significant after an intraoperative 0.5-mg/kg bolus, followed by a 10-mcg/kg/min infusion. In lumbar spinal and scoliosis surgery, studies reported a benefit after a bolus and infusion[56–58]; but a negative finding was reported in pediatric spinal surgery.[59]

To date, there are no data reporting the analgesic effects of ketamine in brain surgery; this may be due to historical association with an increase intracranial pressure (ICP). Clinical investigations into this phenomenon have shown the contrary. In fact, even when used for patients with traumatic brain injury, ICP measurements remained similar to opioid administration, with some studies showing that a ketamine bolus may lower ICP.[60–65]

There exists clinical and experimental evidence that a bolus dose of ketamine and intraoperative infusion augments somatosensory evoked potentials and motor evoked potential (MEP) monitoring.[66,67] Studies have shown improved cortical signal amplitude[68] and that infusions do not increase voltage required to elicit maximum MEP amplitude.[69]

Lidocaine

Lidocaine has recently garnered attention as an analgesic adjunct. A Cochrane review reported low to moderate evidence to support perioperative lidocaine infusion.[70] A 116-patient study demonstrated statistically significant improved pain scores and physical function and a lower 30-day complication rate when undergoing spinal surgery with a perioperative lidocaine infusion versus a placebo infusion.[71] In neurosurgical patients, especially with pathology in the temporal lobe, caution should be exercised as lidocaine can lower the seizure threshold in a dose-dependent manner, which has been shown to start in the amygdala.[72,73]

Corticosteroids

For palliative care, corticosteroids have been used as adjuvant therapy for metastatic bone pain, neuropathic pain, and visceral pain.[74] In the perioperative setting, a meta-analysis conducted by Waldron and colleagues[75] concluded that dexamethasone is associated with a small but statistically significant improvement in pain scores and a 13% reduction in opioid use. De Olivera and colleagues[76] identified a 0.1-mg/kg perioperative dose as the most effective dose in achieving these results. Although evidence from the corticosteroid randomisation after significant head injury (CRASH) trial found the use of steroids for ICP management less favorable,[77] in chronic pain, this may be a reasonable adjunct to care.

Dexmedetomidine

Dexmedetomidine is one of the newer anesthetic agents and may serve as an effective adjuvant in multimodal analgesia practices. A meta-analysis conducted by Schnabel and colleagues[78] concluded that intraoperative dexmedetomidine is associated with lower postoperative pain scores, opioid sparing, and a lower risk for respiratory adverse events. A separate meta-analysis that compiled randomized controlled trials in intracranial procedures demonstrated more stable perioperative hemodynamics, less intraoperative opioid consumption, and fewer postoperative antiemetic requests.[79]

Gabapentin and Pregabalin

Yu and colleagues[80] performed a systemic review and meta-analysis and provided a level I rating supporting gabapentin and pregabalin in lumbar surgery with evidence associated with improved pain levels and lower opioid requirements. Ho and colleagues[81] support these conclusions in a systematic review of randomized controlled trials and added that gabapentin was associated with an increased risk of sedation but less opioid-related side effects, such as vomiting and pruritus. It is important to note that gabapentin and pregabalin are first-line agents for many neuropathic pain states. It is highly likely that patients living with chronic pain are either currently taking this medication or have discontinued it because of adverse events or lack of perceived benefit.

OPIOID MANAGEMENT

Opioids are foundational to an effective multimodal analgesia plan, but in the setting of tolerant patients, a strategic approach to administration is required to maximize benefits and minimize harm. It is important to meet a patient's opioid requirements to avoid acute withdrawal or undertreating pain. With that said, it has been demonstrated in several studies that high opioid doses is associated with greater postoperative pain levels, poorer functional outcomes, higher likelihood of postoperative ileus and a longer hospital length of stay when undergoing surgery.[12,82–84] Although opioid-induced respiratory depression is a concern when providing high doses, the closed claims database only documented preexisting COT in 8% of the 92 documented claims associated with opioid overdose.[85] Although this may ease providers into believing that long-term use leads to a reduction in the ventilatory risk, the administration of other respiratory depressants and the co-occurrence of central and obstructive sleep apnea demand clinical vigilance perioperatively.

To date, there are no controlled trials that guide intraoperative management for patients on COT. Brill and colleagues[86] and Mitra and Sinatra[87] wrote comprehensive reviews of perioperative management strategies for opioid-dependent patients.

Whenever possible, patients should resume their regimen on the day of the operation, either by taking long-acting opioids with small sips of water or continuing the fentanyl patch. It is important to avoid forced-air warming devices directly over fentanyl patch sites as heat can increase transdermal absorption. If patients are unable to take oral medications preoperatively, providers should calculate equianalgesic IV doses (see **Table 2**) and provide half the daily requirements before or on induction, even if regional or neuraxial anesthesia is provided. The dose may be reduced to account for lower oral bioavailability drugs, such as morphine and hydromorphone (20%–35% bioavailability) with less adjustments needed for methadone, oxycodone, hydrocodone, and fentanyl patch (70%–90% bioavailability).[86]

Front-loading opioids by titrating to a respiratory effect to approximate opioid requirements is a practice commonly performed at the author's institution, particularly for spinal surgery. This practice stems from a study performed by Davis and colleagues[88] whereby 20 patients on COT undergoing a multilevel spinal fusion were administered a fentanyl infusion at 2 mcg/kg/min until the respiratory rate decreased to 5 breaths per minute. This practice allowed for the calculation of the serum plasma concentration of 30% of the level associated with respiratory depression and applied this dose calculation postoperatively to set patient controlled analgesia (PCA) settings. Although a protocol of this nature may be impractical clinically, a modification that involves intermittent boluses of hydromorphone, fentanyl, or morphine while patients are conscious to generate a safe, yet effective dose range intraoperatively has value. Anecdotally, after induction of a general anesthetic with this technique, the provider may have to manage temporary hypotension before the surgical stimulus.

Methadone is a viable alternative for a long-acting alternative to opioid management for patients both on and off COT. In one study, a single bolus of 0.2 mg/kg of methadone provided a 50% decrease in pain reports and reduction of postoperative opioid requirements after complex spinal surgery compared with sufentanil infusion.[89] Methadone is associated with QTc prolongation and risk for torsades de pointes at high doses (greater than 60 mg/d), but most pain medicine regimens provide a lower dose; therefore, the risk is not as significant.[90] In addition, the variable pharmacokinetics of methadone can predispose to delayed respiratory depression postoperatively; therefore, it should be carefully selected for opioid-tolerant patients. Patients may also require closer postsurgical ventilatory and oxygenation monitoring.

NONINTRAVENOUS PHARMACOLOGIC TREATMENT OPTIONS

Peripheral and central neural blockade through regional anesthesia techniques provide superior analgesia than systemic medications, and this has been recognized by the ASA acute pain task force.[32] When appropriate, patients with chronic pain should be offered neuraxial or peripheral anesthesia.[91] Although neurosurgery is not classically associated with regional techniques, there are several methods that are reviewed.

NEURAXIAL ANESTHESIA FOR LUMBAR LAMINECTOMY OF DISC SURGERY

Institutional practices typically guide the anesthetic choice between regional or GA for spinal surgery. A 2014 review reported that spinal and even epidural anesthesia for single- or double-level decompressive surgery has reported more stable hemodynamic profile and better pain control compared with GA.[92] Although this may offer distinct advantages for patients with chronic pain undergoing low complexity laminectomies or discectomies, anesthesiologists should coordinate this approach with surgeons.

Scalp Block

The scalp block has been most commonly performed in awake craniotomies before the placement of cranial pins; however, evidence supports this technique as a useful adjunct to care in patients with GA. The scalp block was first reported in 1996 by Pinosky and colleagues[93]; a subsequent review by Ortiz-Cardona and Bendo[94] provides detailed procedural description and figures. The block involves the infiltration of local anesthetic through soft tissue layers to the skull to anesthetize the supraorbital, zygomatico-temporal, auriculo-temporal, greater occipital, and lesser occipital nerves. This block can be performed before pinning or after skin closure and has been associated with an opioid-sparing effect and improved pain scores.[94]

Infiltration

In spinal and cranial surgeries, wound infiltration has also been investigated as an adjunct to a multimodal strategy. Historically, scalp infiltration has been considered inferior to selective scalp block, with minimal postoperative benefit.[94] However, a small single-blinded study by Batoz and colleagues[95] reported that infiltration at the surgical site with ropivacaine can lead to a small decreased risk in the development of postsurgical neuropathic pain.

In lumbar spinal surgery, infiltrating longer-acting local anesthetics like bupivacaine into the incision site have been investigated, with many studies showing an opioid-sparing effect and improved pain scores.[96–100] Some evidence suggests that preincisional infiltration has been shown superior to infiltration during wound closure.[96,97,100] Bone graft harvesting of the anterior superior iliac spine for spinal fusion surgeries is associated with significant postoperative pain, including chronic pain at the harvest site. Continuous infusion catheters with bupivacaine at the bone harvest site have been investigated in randomized, double-blind controlled trials, with one group reporting significant analgesia and reduction in opioid requirements[101] and another showing no effect.[102] It is pertinent to note that liposomal bupivacaine, a compound that provides a sustained release of bupivacaine from multivesicular liposomes, has been approved by the Food and Drug Administration in 2015 for a single-shot infiltration of surgical incisions. Although it has not been trialed in neurosurgical patients, most studies have documented analgesia and opioid sparing lasting up to 72 hours compared with the 8 to 12 hours of relief in conventional bupivacaine.[103]

SUMMARY

Patients with chronic pain undergoing neurosurgery may induce concern in anesthesiologists, especially when considering the delicate nature of the procedure. Yet, this challenge should be seen as an opportunity to impact patient care acutely, subacutely and chronically. It is all too common for patients on COT to undergo a surgical procedure, have poorly treated pain, and then leave the hospital with a higher daily opioid dose that remains long-term. Providing the most comfortable experience possible during the recovery period through a multimodal strategy can have lasting benefits. This point is particularly important in spinal and cranial surgery whereby the immediate rehabilitation efforts are critical to improving long-term outcomes.

In the end, after appraising existent evidence, more research is needed to better define best practices, identify treatment dangers, and risk stratify patients with chronic pain when undergoing neurosurgical procedures. At present, anesthesiologists should become familiar with various analgesic techniques to adapt a personalized multimodal strategy to each unique patient undergoing surgery.

REFERENCES

1. Johannes CB, Le TK, Zhou X, et al. The prevalence of chronic pain in United States adults: results of an Internet-based survey. J Pain 2010;11(11):1230–9.
2. Stockbridge EL, Suzuki S, Pagan JA. Chronic pain and health care spending: an analysis of longitudinal data from the Medical Expenditure Panel Survey. Health Serv Res 2015;50(3):847–70.
3. Centers for Disease Control and Prevention. Vital signs: overdoses of prescription opioid pain relievers–United States, 1999–2008. MMWR Morb Mortal Wkly Rep 2011;60(43):1487–92.
4. Frenk SM, Porter KS, Paulozzi LJ. Prescription opioid analgesic use among adults: United States, 1999-2012. NCHS Data Brief 2015;(189):1–8.
5. Kim KH, Moon SH, Hwang CJ, et al. Prevalence of neuropathic pain in patients scheduled for lumbar spine surgery: nationwide, multicenter, prospective study. Pain Physician 2015;18(5):E889–97.
6. Imamura M, Chen J, Matsubayashi SR, et al. Changes in pressure pain threshold in patients with chronic nonspecific low back pain. Spine 2013; 38(24):2098–107.
7. Younger J, Barelka P, Carroll I, et al. Reduced cold pain tolerance in chronic pain patients following opioid detoxification. Pain Med 2008;9(8):1158–63.
8. Gibson SJ, Littlejohn GO, Gorman MM, et al. Altered heat pain thresholds and cerebral event-related potentials following painful CO_2 laser stimulation in subjects with fibromyalgia syndrome. Pain 1994;58(2):185–93.
9. Kalkman CJ, Visser K, Moen J, et al. Preoperative prediction of severe postoperative pain. Pain 2003;105(3):415–23.
10. Janssen KJ, Vergouwe Y, Kalkman CJ, et al. A simple method to adjust clinical prediction models to local circumstances. Can J Anaetsh 2009;56(3):194–201.
11. Janssen KJ, Kalkman CJ, Grobbee DE, et al. The risk of severe postoperative pain: modification and validation of a clinical prediction rule. Anesth Analg 2008;107(4):1330–9.
12. Chapman CR, Davis J, Donaldson GW, et al. Postoperative pain trajectories in chronic pain patients undergoing surgery: the effects of chronic opioid pharmacotherapy on acute pain. J Pain 2011;12(12):1240–6.
13. Melzack R, Wall PD. Pain mechanisms: a new theory. Science 1965;150(3699): 971–9.
14. Sieweke N, Birklein F, Riedl B, et al. Patterns of hyperalgesia in complex regional pain syndrome. Pain 1999;80(1–2):171–7.
15. Melzack R. Pain and the neuromatrix in the brain. J Dent Educ 2001;65(12): 1378–82.
16. Melzack R. From the gate to the neuromatrix. Pain 1999;(Suppl 6):S121–6.
17. Moseley GL. A pain neuromatrix approach to patients with chronic pain. Man Ther 2003;8(3):130–40.
18. Cagnie B, Coppieters I, Denecker S, et al. Central sensitization in fibromyalgia? A systematic review on structural and functional brain MRI. Semin Arthritis Rheum 2014;44(1):68–75.
19. Peyron R, Laurent B, Garcia-Larrea L. Functional imaging of brain responses to pain. A review and meta-analysis (2000). Neurophysiol Clin 2000;30(5):263–88.
20. Chaparro LE, Smith SA, Moore RA, et al. Pharmacotherapy for the prevention of chronic pain after surgery in adults. Cochrane Database Syst Rev 2013;(7):CD008307.

21. Flexman AM, Ng JL, Gelb AW. Acute and chronic pain following craniotomy. Curr Opin Anaesthesiol 2010;23(5):551–7.
22. McGreevy K, Bottros MM, Raja SN. Preventing chronic pain following acute pain: risk factors, preventive strategies, and their efficacy. Eur J Pain Suppl 2011;5(2):365–72.
23. Sidhu GS, Henkelman E, Vaccaro AR, et al. Minimally invasive versus open posterior lumbar interbody fusion: a systematic review. Clin Orthop Relat Res 2014; 472(6):1792–9.
24. Wang J, Zhou Y, Zhang ZF, et al. Comparison of one-level minimally invasive and open transforaminal lumbar interbody fusion in degenerative and isthmic spondylolisthesis grades 1 and 2. Eur Spine J 2010;19(10):1780–4.
25. Park Y, Ha JW. Comparison of one-level posterior lumbar interbody fusion performed with a minimally invasive approach or a traditional open approach. Spine 2007;32(5):537–43.
26. Sasso RC, LeHuec JC, Shaffrey C, Spine Interbody Research Group. Iliac crest bone graft donor site pain after anterior lumbar interbody fusion: a prospective patient satisfaction outcome assessment. J Spinal Disord Tech 2005;18:S77–81.
27. Gottschalk A, Berkow LC, Stevens RD, et al. Prospective evaluation of pain and analgesic use following major elective intracranial surgery. J Neurosurg 2007; 106(2):210–6.
28. Harner SG, Beatty CW, Ebersold MJ. Impact of cranioplasty on headache after acoustic neuroma removal. Neurosurgery 1995;36(6):1097–9.
29. Koperer H, Deinsberger W, Jodicke A, et al. Postoperative headache after the lateral suboccipital approach: craniotomy versus craniectomy. Minim Invasive Neurosurg 1999;42(4):175–8.
30. Evers AW, Kraaimaat FW, Geenen R, et al. Pain coping and social support as predictors of long-term functional disability and pain in early rheumatoid arthritis. Behav Res Ther 2003;41(11):1295–310.
31. Ang DC, Bair MJ, Damush TM, et al. Predictors of pain outcomes in patients with chronic musculoskeletal pain co-morbid with depression: results from a randomized controlled trial. Pain Med 2010;11(4):482–91.
32. American Society of Anesthesiologists Task Force on Acute Pain Management. Practice guidelines for acute pain management in the perioperative setting: an updated report by the American Society of Anesthesiologists Task Force on Acute Pain Management. Anesthesiology 2012;116(2):248–73.
33. Mathiesen O, Dahl B, Thomsen BA, et al. A comprehensive multimodal pain treatment reduces opioid consumption after multilevel spine surgery. Eur Spine J 2013;22(9):2089–96.
34. Garcia RM, Cassinelli EH, Messerschmitt PJ, et al. A multimodal approach for postoperative pain management after lumbar decompression surgery: a prospective, randomized study. J Spinal Disord Tech 2013;26(6):291–7.
35. Devin CJ, McGirt MJ. Best evidence in multimodal pain management in spine surgery and means of assessing postoperative pain and functional outcomes. J Clin Neurosci 2015;22(6):930–8.
36. Tzortzopoulou A, McNicol ED, Cepeda MS, et al. Single dose intravenous propacetamol or intravenous paracetamol for postoperative pain. Cochrane Database Syst Rev 2011;(10):CD007126.
37. Roberts M, Brodribb W, Mitchell G. Reducing the pain: a systematic review of postdischarge analgesia following elective orthopedic surgery. Pain Med 2012;13(5):711–27.

38. Cassinelli EH, Dean CL, Garcia RM, et al. Ketorolac use for postoperative pain management following lumbar decompression surgery: a prospective, randomized, double-blinded, placebo-controlled trial. Spine 2008;33(12):1313–7.

39. Lamberts M, Lip GY, Hansen ML, et al. Relation of nonsteroidal anti-inflammatory drugs to serious bleeding and thromboembolism risk in patients with atrial fibrillation receiving antithrombotic therapy: a nationwide cohort study. Ann Intern Med 2014;161(10):690–8.

40. Schjerning Olsen AM, Gislason GH, McGettigan P, et al. Association of NSAID use with risk of bleeding and cardiovascular events in patients receiving antithrombotic therapy after myocardial infarction. JAMA 2015;313(8):805–14.

41. Shin JY, Park MJ, Lee SH, et al. Risk of intracranial haemorrhage in antidepressant users with concurrent use of non-steroidal anti-inflammatory drugs: nationwide propensity score matched study. BMJ 2015;351:h3517.

42. Richardson MD, Palmeri NO, Williams SA, et al. Routine perioperative ketorolac administration is not associated with hemorrhage in pediatric neurosurgery patients. J Neurosurg Pediatr 2016;17(1):107–15.

43. Magni G, La Rosa I, Melillo G, et al. Intracranial hemorrhage requiring surgery in neurosurgical patients given ketorolac: a case-control study within a cohort (2001-2010). Anesth Analg 2013;116(2):443–7.

44. Niesters M, Martini C, Dahan A. Ketamine for chronic pain: risks and benefits. Br J Clin Pharmacol 2014;77(2):357–67.

45. Patil S, Anitescu M. Efficacy of outpatient ketamine infusions in refractory chronic pain syndromes: a 5-year retrospective analysis. Pain Med 2012; 13(2):263–9.

46. Azari P, Lindsay DR, Briones D, et al. Efficacy and safety of ketamine in patients with complex regional pain syndrome: a systematic review. CNS Drugs 2012; 26(3):215–28.

47. Kiefer RT, Rohr P, Ploppa A, et al. Efficacy of ketamine in anesthetic dosage for the treatment of refractory complex regional pain syndrome: an open-label phase II study. Pain Med 2008;9(8):1173–201.

48. Kapural L, Kapural M, Bensitel T, et al. Opioid-sparing effect of intravenous outpatient ketamine infusions appears short-lived in chronic-pain patients with high opioid requirements. Pain Physician 2010;13(4):389–94.

49. Bell RF, Eccleston C, Kalso EA. Ketamine as an adjuvant to opioids for cancer pain. Cochrane Database Syst Rev 2012;(11):CD003351.

50. Elia N, Tramer MR. Ketamine and postoperative pain–a quantitative systematic review of randomised trials. Pain 2005;113(1–2):61–70.

51. Carstensen M, Moller AM. Adding ketamine to morphine for intravenous patient-controlled analgesia for acute postoperative pain: a qualitative review of randomized trials. Br J Anaesth 2010;104(4):401–6.

52. Subramaniam K, Subramaniam B, Steinbrook RA. Ketamine as adjuvant analgesic to opioids: a quantitative and qualitative systematic review. Anesth Analg 2004;99(2):482–95, table of contents.

53. Bell RF, Dahl JB, Moore RA, et al. Peri-operative ketamine for acute postoperative pain: a quantitative and qualitative systematic review (Cochrane review). Acta Anaesthesiol Scand 2005;49(10):1405–28.

54. Loftus RW, Yeager MP, Clark JA, et al. Intraoperative ketamine reduces perioperative opiate consumption in opiate-dependent patients with chronic back pain undergoing back surgery. Anesthesiology 2010;113(3):639–46.

55. Dahi-Taleghani M, Fazli B, Ghasemi M, et al. Effect of intravenous patient controlled ketamine analgesia on postoperative pain in opium abusers. Anesth Pain Med 2014;4(1):e14129.
56. Kim SH, Kim SI, Ok SY, et al. Opioid sparing effect of low dose ketamine in patients with intravenous patient-controlled analgesia using fentanyl after lumbar spinal fusion surgery. Korean J Anesthesiol 2013;64(6):524–8.
57. Pacreu S, Fernandez Candil J, Molto L, et al. The perioperative combination of methadone and ketamine reduces post-operative opioid usage compared with methadone alone. Acta Anaesthesiol Scand 2012;56(10):1250–6.
58. Jabbour HJ, Naccache NM, Jawish RJ, et al. Ketamine and magnesium association reduces morphine consumption after scoliosis surgery: prospective randomised double-blind study. Acta Anaesthesiol Scand 2014;58(5):572–9.
59. Pestieau SR, Finkel JC, Junqueira MM, et al. Prolonged perioperative infusion of low-dose ketamine does not alter opioid use after pediatric scoliosis surgery. Paediatr Anaesth 2014;24(6):582–90.
60. Zeiler FA, Teitelbaum J, West M, et al. The ketamine effect on ICP in traumatic brain injury. Neurocrit Care 2014;21(1):163–73.
61. Wang X, Ding X, Tong Y, et al. Ketamine does not increase intracranial pressure compared with opioids: meta-analysis of randomized controlled trials. J Anesth 2014;28(6):821–7.
62. Bourgoin A, Albanese J, Wereszczynski N, et al. Safety of sedation with ketamine in severe head injury patients: comparison with sufentanil. Crit Care Med 2003;31(3):711–7.
63. Albanese J, Arnaud S, Rey M, et al. Ketamine decreases intracranial pressure and electroencephalographic activity in traumatic brain injury patients during propofol sedation. Anesthesiology 1997;87(6):1328–34.
64. Chang LC, Raty SR, Ortiz J, et al. The emerging use of ketamine for anesthesia and sedation in traumatic brain injuries. CNS Neurosci Ther 2013;19(6):390–5.
65. Mayberg TS, Lam AM, Matta BF, et al. Ketamine does not increase cerebral blood flow velocity or intracranial pressure during isoflurane/nitrous oxide anesthesia in patients undergoing craniotomy. Anesth Analg 1995;81(1):84–9.
66. Langeron O, Lille F, Zerhouni O, et al. Comparison of the effects of ketamine-midazolam with those of fentanyl-midazolam on cortical somatosensory evoked potentials during major spine surgery. Br J Anaesth 1997;78(6):701–6.
67. Kochs E, Bischoff P. Ketamine and evoked potentials. Anaesthesist 1994;43: S8–14.
68. Schubert A, Licina MG, Lineberry PJ. The effect of ketamine on human somatosensory evoked potentials and its modification by nitrous oxide. Anesthesiology 1990;72(1):33–9.
69. Zaarour C, Engelhardt T, Strantzas S, et al. Effect of low-dose ketamine on voltage requirement for transcranial electrical motor evoked potentials in children. Spine 2007;32(22):E627–30.
70. Kranke P, Jokinen J, Pace NL, et al. Continuous intravenous perioperative lidocaine infusion for postoperative pain and recovery. Cochrane Database Syst Rev 2015;(7):CD009642.
71. Farag E, Ghobrial M, Sessler DI, et al. Effect of perioperative intravenous lidocaine administration on pain, opioid consumption, and quality of life after complex spine surgery. Anesthesiology 2013;119(4):932–40.
72. DeToledo JC, Minagar A, Lowe MR. Lidocaine-induced seizures in patients with history of epilepsy: effect of antiepileptic drugs. Anesthesiology 2002;97(3): 737–9.

73. DeToledo JC. Lidocaine and seizures. Drug Monit 2000;22(3):320–2.

74. Watanabe S, Bruera E. Corticosteroids as adjuvant analgesics. J Pain Symptom Manage 1994;9(7):442–5.

75. Waldron NH, Jones CA, Gan TJ, et al. Impact of perioperative dexamethasone on postoperative analgesia and side-effects: systematic review and meta-analysis. Br J Anaesth 2013;110(2):191–200.

76. De Oliveira GS Jr, Almeida MD, Benzon HT, et al. Perioperative single dose systemic dexamethasone for postoperative pain: a meta-analysis of randomized controlled trials. Anesthesiology 2011;115(3):575–88.

77. Edwards P, Arango M, Balica L, et al. Final results of MRC CRASH, a randomised placebo-controlled trial of intravenous corticosteroid in adults with head injury-outcomes at 6 months. Lancet 2005;365(9475):1957–9.

78. Schnabel A, Meyer-Friessem CH, Reichl SU, et al. Is intraoperative dexmedetomidine a new option for postoperative pain treatment? A meta-analysis of randomized controlled trials. Pain 2013;154(7):1140–9.

79. Peng K, Wu S, Liu H, et al. Dexmedetomidine as an anesthetic adjuvant for intracranial procedures: meta-analysis of randomized controlled trials. J Clin Neurosci 2014;21(11):1951–8.

80. Yu L, Ran B, Li M, et al. Gabapentin and pregabalin in the management of postoperative pain after lumbar spinal surgery: a systematic review and meta-analysis. Spine 2013;38(22):1947–52.

81. Ho KY, Gan TJ, Habib AS. Gabapentin and postoperative pain–a systematic review of randomized controlled trials. Pain 2006;126(1–3):91–101.

82. Lawrence JT, London N, Bohlman HH, et al. Preoperative narcotic use as a predictor of clinical outcome: results following anterior cervical arthrodesis. Spine 2008;33(19):2074–8.

83. Kessler ER, Shah M, Gruschkus SK, et al. Cost and quality implications of opioid-based postsurgical pain control using administrative claims data from a large health system: opioid-related adverse events and their impact on clinical and economic outcomes. Pharmacotherapy 2013;33(4):383–91.

84. Barletta JF. Clinical and economic burden of opioid use for postsurgical pain: focus on ventilatory impairment and ileus. Pharmacotherapy 2012;32(9 Suppl):12S–8S.

85. Lee LA, Caplan RA, Stephens LS, et al. Postoperative opioid-induced respiratory depression: a closed claims analysis. Anesthesiology 2015;122(3):659–65.

86. Brill S, Ginosar Y, Davidson EM. Perioperative management of chronic pain patients with opioid dependency. Curr Opin Anaesthesiol 2006;19(3):325–31.

87. Mitra S, Sinatra RS. Perioperative management of acute pain in the opioid-dependent patient. Anesthesiology 2004;101(1):212–27.

88. Davis JJ, Swenson JD, Hall RH, et al. Preoperative "fentanyl challenge" as a tool to estimate postoperative opioid dosing in chronic opioid-consuming patients. Anesth Analg 2005;101(2):389–95.

89. Gottschalk A, Durieux ME, Nemergut EC. Intraoperative methadone improves postoperative pain control in patients undergoing complex spine surgery. Anesth Analg 2011;112(1):218–23.

90. Grodofsky S, Edson E, Huang S, et al. The QTc effect of low-dose methadone for chronic pain: a prospective pilot study. Pain Med 2015;16(6):1112–21.

91. Souzdalnitski D, Halaszynski TM, Faclier G. Regional anesthesia and co-existing chronic pain. Curr Opin Anaesthesiol 2010;23(5):662–70.

92. De Rojas JO, Syre P, Welch WC. Regional anesthesia versus general anesthesia for surgery on the lumbar spine: a review of the modern literature. Clin Neurol Neurosurg 2014;119:39–43.
93. Pinosky ML, Fishman RL, Reeves ST, et al. The effect of bupivacaine skull block on the hemodynamic response to craniotomy. Anesth Analg 1996;83(6):1256–61.
94. Ortiz-Cardona J, Bendo AA. Perioperative pain management in the neurosurgical patient. Anesthesiol Clin 2007;25(3):655–74, xi.
95. Batoz H, Verdonck O, Pellerin C, et al. The analgesic properties of scalp infiltrations with ropivacaine after intracranial tumoral resection. Anesth Analg 2009;109(1):240–4.
96. Gurbet A, Bekar A, Bilgin H, et al. Pre-emptive infiltration of levobupivacaine is superior to at-closure administration in lumbar laminectomy patients. Eur Spine J 2008;17(9):1237–41.
97. Gurbet A, Bekar A, Bilgin H, et al. Preemptive wound infiltration in lumbar laminectomy for postoperative pain: comparison of bupivacaine and levobupivacaine. Turk Neurosurg 2014;24(1):48–53.
98. Hernandez-Palazon J, Tortosa Serrano JA, Burguillos Lopez S, et al. Infiltration of the surgical wound with local anesthetic for postoperative analgesia in patients operated on for lumbar disc herniation. Comparative study of ropivacaine and bupivacaine. Rev Esp Anestesiol Reanim 2001;48(1):17–20.
99. Jonnavithula N, Garre S, Pasupuleti S, et al. Wound instillation of local anesthetic bupivacaine for postoperative analgesia following lumbar laminectomy. Middle East J Anaesthesiol 2015;23(2):193–8.
100. Ersayli DT, Gurbet A, Bekar A, et al. Effects of perioperatively administered bupivacaine and bupivacaine-methylprednisolone on pain after lumbar discectomy. Spine 2006;31(19):2221–6.
101. Singh K, Samartzis D, Strom J, et al. A prospective, randomized, double-blind study evaluating the efficacy of postoperative continuous local anesthetic infusion at the iliac crest bone graft site after spinal arthrodesis. Spine 2005;30(22):2477–83.
102. Morgan SJ, Jeray KJ, Saliman LH, et al. Continuous infusion of local anesthetic at iliac crest bone-graft sites for postoperative pain relief. A randomized, double-blind study. J Bone Joint Surg Am. 2006;88(12):2606–12.
103. Chahar P, Cummings KC. Liposomal bupivacaine: a review of a new bupivacaine formulation. J Pain Res 2012;5:257–64.

Anesthesia for Endovascular Approaches to Acute Ischemic Stroke

Rafi Avitsian, MD*, Sandra B. Machado, MD

KEYWORDS

- Ischemic stroke • Endovascular therapy • Anesthetic methods • General anesthesia
- Procedural sedation • Hemodynamic monitoring • Neuroprotection

KEY POINTS

- There is a need for anesthetic management in most of the patients undergoing endovascular therapy of stroke, which includes but is not limited to an anesthetic plan of sedation or general anesthesia; hemodynamic, respiratory, intravascular fluids, glycemic control, and neuroprotection are also essential for a favorable outcome.
- Despite multiple retrospective studies showing a better outcome when general anesthesia is avoided if possible, there is still a need for prospective well-designed studies.
- Currently the Society for Neurosciences in Anesthesiology and Critical Care consensus on the anesthetic management of patients undergoing endovascular treatment is the best available guide.
- Neuroprotection is an evolving topic in neuroanesthesia mostly due to difficulties in translation of animal data to humans. Remote ischemic perconditioning might be a very promising field that needs further investigation.

INTRODUCTION

Despite recent advances in acute ischemic stroke (AIS) treatment, this disease still remains a major contributor to mortality and morbidity. The outcome is entirely dependent on rapid diagnosis and early treatment, namely the time factor. A delay in diagnosis can limit the number of patients eligible for effective treatment to stop progression and reverse the course of its pathophysiology. Although the US Food and Drug Administration has approved recombinant tissue plasminogen activator to be used intravenously up to 3 hours after onset of AIS,[1] only approximately 2% to 5% of patients affected with AIS receive thrombolytic therapy.[2] This approach especially after the publication of European Cooperative Acute Stroke study has been used up to 4 to 5 hours after the initiation of the symptoms but benefited a small

This work is published in collaboration with the Society for Neuroscience in Anesthesiology and Critical Care.
Department of General Anesthesiology, Cleveland Clinic, 9500 Euclid Avenue E-31, Cleveland, OH 44195, USA
* Corresponding author.
E-mail address: avitsir@ccf.org

population.[3,4] Intra-arterial treatment of AIS is a new strategy that is intended to expand the population eligible for treatment, that is, attenuates the time limitation and patient's resistance to intravenous thrombolytic therapy. Currently intra-arterial treatment is being used up to 6 hours after an AIS resulting from middle cerebral artery occlusion, although a futile recanalization is not uncommon especially in an older patient population and those with more severe neurologic deficits.[5] The additional benefit of intra-arterial thrombolysis beyond intravenous thrombolysis is under investigation. A systematic review and meta-analysis[6] show a benefit of intra-arterial thrombolysis over standard therapy but not a clear benefit over intravenous thrombolysis.

In most centers, endovascular treatment requires anesthetic intervention, necessitating a better understanding of the effect of anesthetic technique on the disease process to improve outcome. A team-based approach involving the stroke neurologist, neuroanesthesiologist, neurointerventionalist, and neurocritical specialist is essential. Sharing information regarding current studies with these subspecialties adds a piece to the puzzle of "best acute stroke therapy."

This article summarizes recent advances relevant to neuroanesthesia for interventional treatment of AIS with focus on 3 areas: anesthetic method, hemodynamic management, and brain protecting measures during endovascular treatment.

ANESTHETIC MANAGEMENT

The importance of anesthetic management in patients with AIS is not a matter of debate, but the effect of the anesthetic technique on the success of reperfusion is still a topic for discussion. The literature is rich with nonanesthesiologists commenting on the anesthetic method and hemodynamics during endovascular procedures.[7–10]

One of the most important hurdles in describing the best anesthetic method is the not-so-clear understanding of what general anesthesia (GA) is, especially for nonanesthesiologist researchers.[11] Also there is the variability in anesthetic techniques for the wide range of what is known as GA.[12] There is a gap in well-designed prospective studies that investigate the effect of anesthetic method isolated from confounding factors. Due to different effects of anesthetics and their individual potential for protective or harmful consequences, it is wrong to generalize the effect of one method to all. A recent meta-analysis of 9 studies enrolling 1956 patients showed a worse outcome in patients undergoing GA.[13] The results of this meta-analysis showed that GA had higher odds of mortality, lower odds of favorable functional outcome, but fewer adverse respiratory events. The investigators accepted that the difference within the stroke severity may have been the confounding factor in the result, again stressing the need for a better randomized study.

The debate on the best anesthetic management for intravascular treatment of AIS started fading following the release of the Society for Neurosciences in Anesthesiology and Critical Care (SNACC) Clinical Consensus Statement in 2014.[14] One main reason is the realization by both anesthesia and neurointerventional groups of the paucity of well-designed studies. The published statement was the product of a task force that did not limit itself to a concise literature search. The significance of the task force was the involvement of the Society of Neurointerventional Surgery, as well as the Neurocritical Care Society and soliciting multispecialty input. The recommendations were thus published as an endorsement of SNACC in the *Journal of Neurosurgical Anesthesiology* as well as *Stroke*, making it more acceptable to clinicians on both sides. However, the level of evidence for most recommendations mentioned in the consensus was not very high. Specifically for the anesthetic technique for the endovascular treatment of AIS, the task force points out the selection bias in available literature on the better outcome in using local anesthesia with sedation compared with GA for these procedures. The level of evidence was according to the standard of American Heart Association evidence rating scheme, and regarding the anesthetic method were at best class II with level of evidence B, especially those related to use of GA in uncooperative patients and patients with posterior circulation ischemic strokes. In fact this translates into "additional studies are needed" and evidence is based on single trials and nonrandomized studies. The same level of evidence has been given to the feasibility of using local anesthesia with sedation in patients who are able to control their airway, without the definite consensus that this is the preferred option according to strong evidence. However one relevant and appropriate recommendation related to the anesthetic method is the importance of speed in the management of AIS, which could, perhaps mistakenly, imply that using local anesthesia and sedation can be faster compared with GA. Thus, we see that there is still a lack of adequate evidence showing the superiority of one method of anesthesia over the other.

Even after the consensus statement, other studies have been published investigating the role of anesthetic methods in the outcome after treatment of AIS.[15] Although the studies were better designed with an investigation focusing on confounding variables and improvement in a more scientific inference, there is still a lack of hard evidence and a need for larger prospective randomized multicenter studies to evaluate the effect of anesthetic technique.[16] The calculation of the size of a study that has enough power to show a benefit of one method over the other is very complex considering all the many confounding factors in addition to a large variable patient population. A very important determinant of outcome is the National Institutes of Health Stroke Scale (NIHSS) score at presentation time. Comorbidities in patients with AIS are also an important element of the outcome regardless of the extent of ischemic area, timing, and presence or absence of initial thrombolytic therapy. Thus, the NIHSS subscores or other measures of illness severity that encompass non-neurologic factors may be better prognostic identifiers.[17]

The class of anesthetic agents used and their doses also may have an impact on level of brain protection or harm, which is discussed later in this article. Moreover, one cannot underestimate the importance of blood pressure (BP) and its deviation from patient's baseline, and $Paco_2$ and temperature during the treatment period and beyond in the outcome. Thus, to obtain reliable data, a large sample size may be needed in future studies so as to be able to account for each of these and other confounding variables. This scientific rigor could be a daunting challenge for investigators. This difficult study design is in contrast with the flawed current information and necessary reliance on common sense, or better yet theoretic consideration in generalizing some of the established observations in AIS patient management. A very good example is the knowledge about hypotension and its deleterious effect on stroke outcome. Thus, any intervention that may lower BP should be discouraged. Regardless of evidence, it is universally accepted that induction of GA in a patient with a history of hypertension, along with altered cerebral autoregulation and some degree of dehydration, plus a partially depleted sympathetic drive has a higher chance of hypotension. In summary, there is an overall assumption that GA is the culprit for poor outcome but in reality it is most likely related to the hypotension that might happen during induction.

Another often-stated inference is that induction of GA may delay the surgical procedure. The anesthesiologist should take a leadership role and help coordinate the different teams to work simultaneously and in parallel so as to expedite this process; that is, start the preparation and draping of the patient in parallel to the induction process. Large institutions with sizable anesthesia departments may have a higher rate of variations in practice, but with dedicated subspecialty anesthesia sections where anesthesia and procedure personnel have a higher rate of interaction, fewer delays are expected in the workflow. This may be as a result of a dedicated neuroanesthesiology subspecialist attending to these cases during as well as out of regular hours, or protocols prepared by this subspecialist group for the general anesthesiologist. Implementing checklists and protocols for proceduralist and anesthesiologist can decrease variability. Preparation and access for intervention can start parallel to but should not be disruptive of a safe anesthetic induction. The nursing team should be available to assist the anesthesiologist if needed. A potential for delay can be the time required to start an invasive BP monitoring. Again this could be done in parallel with the procedure or by getting a side line from the access catheter placed by the proceduralist until a different dedicated arterial line is established. In situations in which GA is necessary in an unstable patient to the point that initiation of an arterial monitoring device is essential even before induction of anesthesia, a team-based approach with involvement of proceduralist is recommended.

For quality improvement purposes, definition of delay and time to readiness is also important.[11] The definition should be well established, as one can interpret it as when the anesthesia team gets the patient ready for the procedure versus time to vascular access or revascularization, the latter of which is affected by the occluded vessel involved and the interventionalist experience. However, there seems to be a consensus that use of GA prolongs patient readiness for revascularization.[18] But there can be a higher chance of patient movement if patient is not under GA, making the procedure more technically challenging[19] and a longer time to revascularization. Procedure time itself is a determinant of outcome with longer procedure times leading to worse outcome.[20]

A review of patient medication is also important in determining an anesthetic plan and course. A special attention should be given to antihypertensive therapy, discussed later in this article, and antiplatelet or anticoagulant medications, including prior thrombolytic. In all emergent surgical cases one is concerned about the latter medications because they can increase the possibility of intraoperative or postoperative hemorrhage or complicate intra-arterial BP monitoring. In a recent study including a patient registry, a multivariate analysis did not show a detrimental outcome for patients on antiplatelet therapy as far as increasing the number of symptomatic intracranial hemorrhages.[21] Prior use of statins is also a subject of interest. A review of registry results did not show an improved outcome in patients who were on statins before the stroke and the meta-analysis of prior data[22] showed no benefit or detrimental effect. A higher dose of statin, however, may be responsible for an increase in symptomatic intracranial hemorrhage, as shown in another study, although it still showed an improvement in 3 months.[23] A European cohort study did not show higher 30-day mortality in patients who before admission for AIS were on calcium channel or beta blockers.[24] Another study in the same population showed that current preadmission use of angiotensin-converting enzyme inhibitors or angiotensin receptor antagonists reduced 30-day mortality after stroke.[25]

HEMODYNAMIC MEASURES

The importance of having adequate perfusion to the brain after an ischemic stroke is well-known. More than 60% of patients will have systolic BP (SBP) of more than 160 mm Hg during the first hours of AIS.[1] An SBP of more than 185 mm Hg or a diastolic BP (DBP) of more than 105 mm Hg is contraindicated for intravenous tissue plasminogen activator (tPA) and should be immediately addressed. If tPA is not administered, an SBP of more than 220 mm Hg or a DBP of more than 120 mm Hg should be treated.

Hypotension is also a risk factor for poor neurologic outcomes after AIS.[26] As mentioned previously, a shortcoming of GA is a higher chance of hypotension because there is a sudden decrease in sympathetic activity as well as redistribution of blood volume after induction of anesthesia and initiation of positive pressure ventilation. Therefore, the usual physiologic response to ischemic stroke, which is an increase in BP to ensure adequate perfusion from collaterals, is lost. Multiple retrospective studies review the BP during this procedure showing lower BPs during GA when compared with sedation.[10,27] In another study, although the maximum BP was higher, and the minimum BP was lower in the GA group, this did not show a statistically significant difference despite measurement of hypotensive episodes and number of times a vasopressor was given; however, they did not measure the number of vasopressors as well as the total dose.[18] Löwhagen Hendén and colleagues[28] reported that a drop of greater than 40% from baseline in mean arterial pressure resulted in a bad outcome. However, it is not only the drop in the BP that is important, but its variability. BP variability, according to a recent study, has been independently shown to be associated with the development of neurologic deterioration in the early stage of AIS.[29] This underscores the importance of avoiding BP variation, which is challenging during GA. A recent study has shown an association between BP variability and hemorrhagic transformation in the early stages of intravenous thrombolysis.[30]

The SNACC consensus statement has recommended hemodynamic monitoring as soon as AIS is diagnosed with specific goals of SBP of greater than 140 mm Hg using fluids and vasopressors and less than 180 mm Hg and DBP less than 105 mm Hg. Monitoring of BP is recommended to be continuous or at least every 3 minutes. There is no consensus on the type of vasopressor; however, one should use judgment on the amount of fluids in accordance with the patient's history and cardiac function so as to maintain euvolemia. The BP at the end of the procedure also should be closely monitored because it correlates with the success rate of revascularization and complications after the procedure. However, there is no guideline on where to maintain the BP after the reperfusion. A high BP may increase the chance of hemorrhagic transformation and a low BP may cause hypoperfusion or reocclusion, especially in cases in which the stroke is a result of in situ thrombosis. In many cases, the original BP before the occurrence of ischemic event is not known and even if it is, it may not be the ideal BP for the patient. Communication with the proceduralist is important to determine the best postprocedure BP in these patients.[14]

Attention to oxygenation and ventilation is also an important component of the anesthetic, and is an element of interpreting hemodynamic data. Despite its obvious nature, it is worth mentioning that avoiding hypoxemia is important in management of these patients. The SNACC consensus recommends administering oxygen to keep the saturation above 92% and Pao_2 more than 60 mm Hg; it also recommends avoiding respiratory depression–induced hypercarbia.[14] Hyperoxia has theoretic negative interaction after ischemic stroke[31] but should be studied with reperfusion. Hyperventilation on the other hand can cause vasoconstriction and decreased blood flow, although this effect has yet to be studied in patients with ischemic stroke, the deleterious effect has at least been shown in patients with traumatic brain injury.[32] Thus, in intubated patients, special attention should be given to minute ventilation and $Paco_2$ levels. End-tidal CO_2 level may not be the best measure because there could be a large discrepancy in patients with chronic CO_2 retention, but it is nonetheless an essential monitor. A study comparing the end-tidal CO_2 level of patients undergoing endovascular therapy has shown hyperventilation in patients under conscious sedation with median end-tidal level of 26.57 mm Hg, but this could be because of inaccurate measurement of end-tidal CO_2 in patients with spontaneous ventilation.[33]

Interestingly, this study has not mentioned the respiratory rate difference and corresponding $Paco_2$, making the claim of hyperventilation unreliable.

NEUROPROTECTION

The basic goals for stroke treatment strategies are to restore perfusion as quickly as possible through mechanical or pharmacologic thrombolysis and minimize detrimental effects of reperfusion injury. There is a current need for adjunctive therapies that can have a protective effect on brain against ischemia or to increase collateral blood flow so as to extend the time window before permanent injury occurs. The search for an anesthetic that would offer a neuroprotective mechanism has been reported since Goldstein, Wells and Keats[34] described barbiturate to have such effect. Although many anesthetics are thought to have a neuroprotective effect, no clinical study has to date demonstrated that neuroprotection is feasible in humans. Hence, this topic warrants discussion.

The biochemical pathways involved in neuronal death during ischemia include excitotoxicity, oxidative stress and peri-infarct depolarization followed by inflammation and apoptosis.[35] The lack of perfusion and consequent ATP depletion causes a reduction in energy-dependent processes while simultaneously promoting adverse pathways. As a result, there is an increase in cytosolic calcium and sodium that is further enhanced by glutamate-mediated activation of N-methyl-D-aspartate (NMDA) and AMPA receptors. Increased levels of cytosolic calcium contribute to cell apoptosis. Perfusion restoration will cause further damage through excessive production of superoxide free radicals and other damaging processes largely related to endothelial and blood-brain barrier dysfunction.[36]

Neuroprotection is a broad term that can be defined as promotion of neuronal survival. It can be mediated by methods that decrease neuronal metabolic demand, maintain blood flow in the penumbra area, or minimize the secondary effects of substances released by cell death. Common neuroprotective approaches include pharmacologic, hemodilution, hypothermia, hyperoxia, and ischemic conditioning. Pharmacologic neuroprotection has been a recurrent theme of interest for neuroanesthesiologists, mostly because of encouraging animal studies. More than 23,000 publications discuss stroke and its treatment in various animal models and approximately 2700 clinical trials are listed on the Internet Stroke Trials Registry based on this extensive volume of preclinical work. Although a strong body of literature exists supporting in vivo and in vitro strategies,[37,38] most clinical trials have failed to reproduce these data in humans. This is particularly true for NMDA receptor antagonists, sodium channel blockers, γ-aminobutyric acid agonists, calcium channel blockers, lipid peroxidation inhibitors, and intercellular adhesion molecule (ICAM-1) antibodies, just to name a few. Many explanations have been offered and thoroughly debated to justify this discrepancy.[39–41] Donnan[42] emphasized the need for scientific rigor to achieve valuable clinical data or alternatively the adoption of new strategies. Ginsberg[43] attributed the failure of human translation to a few factors: study variation in time window to treat; absence of solid preclinical data and rigorous experimental design, lack of efficacious drug plasma concentration, and poor follow-up. The Stroke Therapy Academic Industry Roundtable (STAIR) VII[44] addressed these issues with some recommendations for future trials. Importantly, this group suggested that stroke therapy and neuroprotective agents should be administered in a timely fashion. The route of administration also should be considered to achieve this goal. They also highlighted that it is advisable to focus on treatment strategies with multiple and/or combined mechanism of action so as to achieve broader effects.

This multimodal strategy has been extensively discussed and identified as a promising approach in stroke management.[41,45,46]

Induced hypothermia is a widely accepted therapy for patients with cardiac arrest and children with hypoxic ischemic encephalopathy,[47] but its efficacy has not yet been demonstrated as a reliable neuroprotective strategy in clinical trials after AIS. There are multiple neuroprotective mechanisms of hypothermia. It decreases the metabolic rate, reduces oxygen demand, preserves energy stores, and decreases lactate production. Zhang and colleagues[48] reported a modulation of apoptosis mediated by a reduction in proapoptotic factors and increasing antiapoptotic proteins. Hypothermia also has an impact on the inflammatory pathway with a net anti-inflammatory effect,[49] overall suppression of excitatory amino acids release,[50] and preservation of blood-brain barrier integrity.[51] Although physical hypothermia seems like a very attractive adjunct therapy for patients with AIS, there are multiple factors that can affect its efficacy, including timing of initiation, duration, technique to achieve adequate temperature, rewarming speed, and return of cerebral perfusion. There are also some limitations to hypothermia related to side effects, such as hemodynamic instability, arrhythmias, coagulopathy, and shivering. Some of these limitations might be ignored in the laboratory setting but should be addressed before clinical translation. Combination of hypothermia with other neuroprotective therapies also should be considered and further investigated.[49]

Remote ischemic conditioning has been extensively studied in the past few years as a promising approach for neuroprotection in AIS. The use of remote sublethal ischemic stimulation to a different organ before an ischemic event is known as remote ischemic preconditioning (RIPrC). This method has limited application in the treatment of AIS due to the unpredictable nature of the disease, although it may be a feasible in patients with prior transient ischemic episodes who are at risk for imminent strokes. Additionally, ischemic stimulation of a limb delivered during ischemia and before reperfusion is defined as remote ischemic perconditioning (RIPerC). Hahn and colleagues[52] showed that RIPerC and RIPrC are effective in reducing cerebral infarct size, with superior results for neuroprotection in RIPerC comparatively. Multiple studies in young male rodents were reviewed by Hess and colleagues[53] reporting an overall reduction of the infarct size. The investigators also made special remarks on RIPerC timing in relation to reperfusion with a lower efficacy of the method when paired with reperfusion. This therapeutic approach was also proven to be effective in female ovariectomized mice with and without intravenous-tPA at 4 hours.[54] The importance of this model consists in the inclusion of higher-risk patients for stroke, such as postmenopausal women who could benefit from RIPerC therapy.

The suggested molecular mechanism of RIPerC is probably related to remote transfer of protection through humoral and neurogenic pathways partially or alterations in genomic regulation of transcription and translation of neuroprotective/toxic proteins. After brief induction of limb ischemia–reperfusion, there is a release of autacoids, such as adenosine, bradykinin, or opioids,[55–57] in response to local muscle ischemia. These substances reach the brain through the circulation and bind receptors in the cerebral endothelium causing ischemic neuroprotection. Opioid receptors ultimately can activate the AKT signaling pathway that mediates responses, such as cell survival, growth, proliferation, cell migration, and angiogenesis. The activation of the parasympathetic nervous system may increase cerebral blood flow,[58] and may also have an impact by decreasing ischemic injury through the cholinergic anti-inflammatory pathway.[59]

Hougaard and colleagues[60] compared patients treated with tPA followed by mechanical thrombectomy with or without RIPerC. Follow-up MRI and NIHSS were performed after 24 hours and 1 month and a clinical examination was performed after 3 months. The investigators concluded that RIPerC during transportation to the hospital had no statistically significant effect on salvage, infarct size, or infarct progression as measured by MRI in a subgroup of patients and 3-month clinical outcome. On adjustment of data to baseline severity of hypoperfusion, there was an increase in tissue survival after 1 month, suggesting a possible neuroprotective role. Future studies need to address the sites and number of limbs to be conditioned, stimulation duration and timing, and possible association of perconditioning and postconditioning.

Volatile anesthetics also have been suggested for neuroprotection in AIS. Their mechanism is most likely related to reduction in glutamate excitotoxicity and, possibly, opening of potassium channels.[61] Promising preclinical data show beneficial effects of isoflurane in ischemic preconditioning[62,63] and postconditioning.[64,65] These findings may have a major impact in intraoperative management of patients with AIS. Intravenous gabaergic anesthetics, such as thiopental and propofol, also have been considered as neuroprotective agents based on global brain ischemia[66] and temporary focal brain ischemia[67] in nonhuman primate studies. As for other approaches, the results have not successfully translated to humans. A recent retrospective study of endovascular management of AIS suggests superiority of volatile anesthetics over intravenous agents, but results need to be validated by prospective clinical trials.[12]

GLYCEMIC CONTROL

The occurrence of hyperglycemia and hypoglycemia are known to be associated with increased mortality and poor recovery after AIS. Suggested association mechanisms between hyperglycemia and cerebral injury exacerbation include increased oxidative load, blood-brain barrier disruption, cerebral edema, hemorrhagic transformation due to impairment of vascular reaction, and inflammation.[68,69] These data have led to many studies investigating the impact of intensive hyperglycemic control on patients with stroke as a neuroprotective strategy. Unfortunately, currently available evidence-based data fail to identify any clinical benefit of this approach,[70] mostly due to the high risk of hypoglycemia.[71,72]

The updated Cochrane review of 11 randomized controlled trials reiterates the lack of advantage in maintaining serum glucose within lower ranges in the first hours of AIS, reporting no impact on functional outcome, death, or improvement in final neurologic deficit. It was again demonstrated a higher incidence of hypoglycemic episodes in that group.[73] Interestingly, experimental investigation of glucose metabolism in a focal ischemia model with preserved collaterals has demonstrated hypermetabolism in the ischemic penumbra. Arnberg and colleagues[74] speculate that these would be a "physiologic" response to increased energy demands questioning even further tight glucose control in the setting of AIS. There is ample discussion around the impact of hyperglycemia on patients with stroke with or without prior history of diabetes. Capes and colleagues[75] reported that patients without diabetes with acute stroke are more negatively affected by high admission glucose, including increased risk of in-hospital mortality and poor functional recovery. Cochrane[73] subgroup analysis was unable to demonstrate such difference in outcome at 30 or 90 days.

The SNACC consensus statement does have a recommendation on the glycemic control, which includes obtaining baseline serum glucose of patients with AIS. As a level II evidence, it recommends hourly glucose checks and preferentially a protocol-driven insulin treatment of blood sugars in excess of 140 mg/dL, serum glucose maintenance of 70 to 140 mg/dL, and treatment of hypoglycemia for initiated for levels less than 50 mg/dL.[14]

COMPLICATIONS OF INTRAVASCULAR TREATMENT OF ACUTE ISCHEMIC STROKE

Complications during intravascular treatment of AIS is an area that needs more investigation.[76] Multiple studies have shown the impact of the anesthetic technique in the size of infarction, rate of hemorrhagic transformation, and pneumonia just to name a few. Recently a rat model showed lessening of infarction volume and intracranial hemorrhage after isoflurane treatment in tPA exaggerated brain injury.[64] In the same context, intravenous pentobarbital was more protective in a rat ischemia model compared with isoflurane.[77] In a human study, intraparenchymal hemorrhage was more frequent in patients who underwent GA for intravascular treatment of AIS.[18] Regardless of the type of procedure or surgery, there are certain complications attributed to GA. Intubated patients have a higher incidence of pneumonia. This is usually a consequence of aspiration during induction or a delay in extubation due to complications of the procedure or inability to control the airway.[78] The severity and location of the AIS also play a major role in the need of airway protection. Intubation is a necessity in patients that aspirated or are at risk for aspiration as well as patients with inadequate oxygenation or ventilation; however, elective intubation to fulfill the plan of GA may contribute to a worse outcome.

Inadvertent rupture of an intracranial artery is a dire complication that is recognized with extravasation of the dye during angiography as well as sudden hemodynamic changes consistent with increased intracranial pressure. Close communication with the proceduralist regarding need for reversal of heparin with protamine is recommended. Rapid control of hemodynamic changes and need for intubation and hyperventilation in patients under sedation is necessary to reduce chance of increased intracranial pressure following intracranial or subarachnoid hemorrhage.

Rarely, dissection or puncture of systemic arteries may arise anywhere from the point of vascular access to the occluded artery being treated. In addition, ischemia distal to the point of vascular access also may arise. Ongoing monitoring for these complications is required throughout the endovascular procedure.

REFERENCES

1. Adams HP Jr, del Zoppo G, Alberts MJ, et al. Guidelines for the early management of adults with ischemic stroke: a guideline from the American Heart Association/American Stroke Association Stroke Council, Clinical Cardiology Council, Cardiovascular Radiology and Intervention Council, and the Atherosclerotic Peripheral Vascular Disease and Quality of Care Outcomes in Research Interdisciplinary Working Groups: The American Academy of Neurology affirms the value of this guideline as an educational tool for neurologists. Circulation 2007;115(20):e478–534.
2. Walter S, Kostopoulos P, Haass A, et al. Diagnosis and treatment of patients with stroke in a mobile stroke unit versus in hospital: a randomised controlled trial. Lancet Neurol 2012;11(5):397–404.
3. Bluhmki E, Chamorro A, Dávalos A, et al. Stroke treatment with alteplase given 3.0-4.5 h after onset of acute ischaemic stroke (ECASS III): additional outcomes and subgroup analysis of a randomised controlled trial. Lancet Neurol 2009; 8(12):1095–102.
4. la Rosa FD, Khoury J, Kissela BM, et al. Eligibility for intravenous recombinant tissue-type plasminogen activator within a population: The Effect of the European Cooperative Acute Stroke Study (ECASS) III Trial. Stroke 2012;43(6):1591–5.

5. Hussein HM, Georgiadis AL, Vazquez G, et al. Occurrence and predictors of futile recanalization following endovascular treatment among patients with acute ischemic stroke: a multicenter study. Am J Neuroradiology 2010;31(3):454–8.

6. Nam J, Jing H, O'Reilly D. Intra-arterial thrombolysis vs. standard treatment or intravenous thrombolysis in adults with acute ischemic stroke: a systematic review and meta-analysis. Int J Stroke 2015;10(1):13–22.

7. Gupta R. Local is better than general anesthesia during endovascular acute stroke interventions. Stroke 2010;41(11):2718–9.

8. Abou-Chebl A, Lin R, Hussain MS, et al. Conscious sedation versus general anesthesia during endovascular therapy for acute anterior circulation stroke: preliminary results from a retrospective, multicenter study. Stroke 2010;41(6): 1175–9.

9. Jumaa MA, Zhang F, Ruiz-Ares G, et al. Comparison of safety and clinical and radiographic outcomes in endovascular acute stroke therapy for proximal middle cerebral artery occlusion with intubation and general anesthesia versus the non-intubated state. Stroke 2010;41(6):1180–4.

10. Davis MJ, Menon BK, Baghirzada LB, et al. Anesthetic management and outcome in patients during endovascular therapy for acute stroke. Anesthesiology 2012;116(2):396–405.

11. Avitsian R, Somal J. Anesthetic management for intra-arterial therapy in stroke. Curr Opin Anaesthesiol 2012;25(5):523–32.

12. Sivasankar C, Stiefel M, Miano TA, et al. Anesthetic variation and potential impact of anesthetics used during endovascular management of acute ischemic stroke. J Neurointerv Surg 2015. [Epub ahead of print].

13. Brinjikji W, Murad MH, Rabinstein AA, et al. Conscious sedation versus general anesthesia during endovascular acute ischemic stroke treatment: a systematic review and meta-analysis. AJNR Am J Neuroradiol 2015;36(3):525–9.

14. Talke PO, Sharma D, Heyer EJ, et al. Society for Neuroscience in Anesthesiology and Critical Care Expert Consensus Statement: anesthetic management of endovascular treatment for acute ischemic stroke endorsed by the Society of Neuro-Interventional Surgery and the Neurocritical Care Society. J Neurosurg Anesthesiology 2014;26(2):95–108.

15. Anastasian ZH. Anaesthetic management of the patient with acute ischaemic stroke. Br J Anaesth 2014;113(Suppl 2):ii9–16.

16. Li F, Deshaies EM, Singla A, et al. Impact of anesthesia on mortality during endovascular clot removal for acute ischemic stroke. J Neurosurg Anesthesiol 2014; 26(4):286–90.

17. Abdul-Rahim AH, Fulton RL, Sucharew H, et al. National Institutes of Health Stroke Scale Item Profiles as predictor of patient outcome: external validation on independent trial data. Stroke 2015;46:395–400 [Erratum appears in Stroke 2015;46(5):E128].

18. John S, Thebo U, Gomes J, et al. Intra-arterial therapy for acute ischemic stroke under general anesthesia versus monitored anesthesia care. Cerebrovasc Dis 2014;38(4):262–7.

19. Brekenfeld C, Mattle HP, Schroth G. General is better than local anesthesia during endovascular procedures. Stroke 2010;41(11):2716–7.

20. Hassan AE, Chaudhry SA, Miley JT, et al. Microcatheter to recanalization (procedure time) predicts outcomes in endovascular treatment in patients with acute ischemic stroke: when do we stop? AJNR Am J Neuroradiol 2013;34(2):354–9.

21. Meseguer E, Labreuche J, Guidoux C, et al. Outcomes after stroke thrombolysis according to prior antiplatelet use. Int J Stroke 2015;10(2):163–9.

22. Meseguer E, Mazighi M, Lapergue B, et al. Outcomes after thrombolysis in AIS according to prior statin use: a registry and review. Neurology 2012;79(17): 1817–23.
23. Scheitz JF, Seiffge DJ, Tütüncü S, et al. Dose-related effects of statins on symptomatic intracerebral hemorrhage and outcome after thrombolysis for ischemic stroke. Stroke 2014;45(2):509–14.
24. Sundboll J, Schmidt M, Horváth-Puhó E, et al. Impact of preadmission treatment with calcium channel blockers or beta blockers on short-term mortality after stroke: a nationwide cohort study. BMC Neurol 2015;15:24.
25. Sundboll J, Schmidt M, Horváth-Puhó E, et al. Preadmission use of ACE inhibitors or angiotensin receptor blockers and short-term mortality after stroke. J Neurol Neurosurg Psychiatry 2015;86(7):748–54.
26. Castillo J, Leira R, García MM, et al. Blood pressure decrease during the acute phase of ischemic stroke is associated with brain injury and poor stroke outcome. Stroke 2004;35(2):520–6.
27. Jagani M, Brinjikji W, Rabinstein AA, et al. Hemodynamics during anesthesia for intra-arterial therapy of acute ischemic stroke. J Neurointerv Surg 2015. [Epub ahead of print].
28. Löwhagen Hendén P, Rentzos A, Karlsson JE, et al. Hypotension during endovascular treatment of ischemic stroke is a risk factor for poor neurological outcome. Stroke 2015;46(9):2678–80.
29. Chung JW, Kim N, Kang J, et al. Blood pressure variability and the development of early neurological deterioration following acute ischemic stroke. J Hypertens 2015;33(10):2099–106.
30. Liu K, Yan S, Zhang S, et al. Systolic blood pressure variability is associated with severe hemorrhagic transformation in the early stage after thrombolysis. Transl Stroke Res 2016. [Epub ahead of print].
31. Rincon F, Kang J, Maltenfort M, et al. Association between hyperoxia and mortality after stroke: a multicenter cohort study. Crit Care Med 2014;42(2):387–96.
32. Coles JP, Fryer TD, Coleman MR, et al. Hyperventilation following head injury: effect on ischemic burden and cerebral oxidative metabolism. Crit Care Med 2007;35(2):568–78.
33. Mundiyanapurath S, Schönenberger S, Rosales ML, et al. Circulatory and respiratory parameters during acute endovascular stroke therapy in conscious sedation or general anesthesia. J Stroke Cerebrovasc Dis 2015;24(6):1244–9.
34. Goldstein A Jr, Wells BA, Keats AS. Increased tolerance to cerebral anoxia by pentobarbital. Archives internationales de pharmacodynamie et de therapie 1966;161(1):138–43.
35. Mantz J, Degos V, Laigle C. Recent advances in pharmacologic neuroprotection. Eur J Anaesthesiol 2010;27(1):6–10.
36. White BC, Sullivan JM, DeGracia DJ, et al. Brain ischemia and reperfusion: molecular mechanisms of neuronal injury. J Neurol Sci 2000;179(S 1–2):1–33.
37. Green AR. Pharmacological approaches to acute ischaemic stroke: reperfusion certainly, neuroprotection possibly. Br J Pharmacol 2008;153:S325–38.
38. Durukan A, Tatlisumak T. Acute ischemic stroke: overview of major experimental rodent models, pathophysiology, and therapy of focal cerebral ischemia. Pharmacol Biochem Behav 2007;87(1):179–97.
39. Gisvold SE, Sterz F, Abramson NS, et al. Cerebral resuscitation from cardiac arrest: treatment potentials. Crit Care Med 1996;24(2 Suppl):S69–80.
40. Xing C, Arai K, Lo EH, et al. Pathophysiologic cascades in ischemic stroke. Int J Stroke 2012;7(5):378–85.

41. Kofke WA. Incrementally applied multifaceted therapeutic bundles in neuropro-tection clinical trials...time for change. Neurocrit Care 2010;12(3):438–44.
42. Donnan GA. The 2007 Feinberg lecture: a new road map for neuroprotection. Stroke 2008;39(1):242.
43. Ginsberg MD. Neuroprotection for ischemic stroke: past, present and future. Neuropharmacology 2008;55(3):363–89.
44. Albers GW, Goldstein LB, Hess DC, et al. Stroke Treatment Academic Industry Roundtable (STAIR) recommendations for maximizing the use of intravenous thrombolytics and expanding treatment options with intra-arterial and neuropro-tective therapies. Stroke 2011;42(9):2645–50.
45. Rogalewski A, Schneider A, Ringelstein EB, et al. Toward a multimodal neuropro-tective treatment of stroke. Stroke 2006;37(4):1129–36.
46. O'Collins VE, Macleod MR, Donnan GA, et al. Evaluation of combination therapy in animal models of cerebral ischemia. J Cereb Blood Flow Metab 2012;32(4):585–97.
47. Nagel S, Papadakis M, Hoyte L, et al. Therapeutic hypothermia in experimental models of focal and global cerebral ischemia and intracerebral hemorrhage. Expert Rev Neurother 2008;8(8):1255–68.
48. Zhang Z, Sobel RA, Cheng D, et al. Mild hypothermia increases Bcl-2 protein expression following global cerebral ischemia. Brain Res Mol Brain Res 2001; 95(1–2):75–85.
49. Han Z, Liu X, Luo Y, et al. Therapeutic hypothermia for stroke: where to go? Exp Neurol 2015;272:67–77.
50. Nakashima K, Todd MM. Effects of hypothermia on the rate of excitatory amino acid release after ischemic depolarization. Stroke 1996;27(5):913–8.
51. Lee JE, Yoon YJ, Moseley ME, et al. Reduction in levels of matrix metalloprotei-nases and increased expression of tissue inhibitor of metalloproteinase-2 in response to mild hypothermia therapy in experimental stroke. J Neurosurg 2005;103(2):289–97.
52. Hahn CD, Manlhiot C, Schmidt MR, et al. Remote ischemic per-conditioning: a novel therapy for acute stroke? Stroke 2011;42(10):2960–2.
53. Hess DC, Hoda MN, Bhatia K. Remote limb preconditioning and postconditioning will it translate into a promising treatment for acute stroke? Stroke 2013;44(4): 1191–7.
54. Hoda MN, Bhatia K, Hafez SS, et al. Remote ischemic perconditioning is effective after embolic stroke in ovariectomized female mice. Transl Stroke Res 2014;5(4): 484–90.
55. Zhou YL, Fathali N, Lekic T, et al. Remote limb ischemic postconditioning protects against neonatal hypoxic-ischemic brain injury in rat pups by the opioid receptor/akt pathway. Stroke 2011;42(2):439–44.
56. Schoemaker RG, van Heijningen CL. Bradykinin mediates cardiac precondition-ing at a distance. Am J Physiol Heart Circ Physiol 2000;278(5):H1571–6.
57. Hu S, Dong H, Zhang H, et al. Noninvasive limb remote ischemic preconditioning contributes neuroprotective effects via activation of adenosine A1 receptor and redox status after transient focal cerebral ischemia in rats. Brain Res 2012; 1459:81–90.
58. Hoyte LC, Papadakis M, Barber PA, et al. Improved regional cerebral blood flow is important for the protection seen in a mouse model of late phase ischemic pre-conditioning. Brain Res 2006;1121(1):231–7.
59. Tracey KJ. Physiology and immunology of the cholinergic antiinflammatory pathway. J Clin Invest 2007;117(2):289–96.

60. Hougaard KD, Hjort N, Zeidler D, et al. Remote ischemic perconditioning as an adjunct therapy to thrombolysis in patients with acute ischemic stroke: a randomized trial. Stroke 2014;45(1):159–67.
61. Zhang J, Zhou W, Qiao H. Bioenergetic homeostasis decides neuroprotection or neurotoxicity induced by volatile anesthetics: a uniform mechanism of dual effects. Med Hypotheses 2011;77(2):223–9.
62. Sun M, Deng B, Zhao X, et al. Isoflurane preconditioning provides neuroprotection against stroke by regulating the expression of the TLR4 signalling pathway to alleviate microglial activation. Sci Rep 2015;5:1445.
63. Xiang HF, Cao DH, Yang YQ, et al. Isoflurane protects against injury caused by deprivation of oxygen and glucose in microglia through regulation of the Toll-like receptor 4 pathway. J Mol Neurosci 2014;54(4):664–70.
64. Kim EJ, Kim SY, Lee JH, et al. Effect of isoflurane post-treatment on tPA-exaggerated brain injury in a rat ischemic stroke model. Korean J Anesthesiol 2015;68(3):281–6.
65. Lee JJ, Li L, Jung HH, et al. Postconditioning with isoflurane reduced ischemia-induced brain injury in rats. Anesthesiology 2008;108(6):1055–62.
66. Bleyaert AL, Nemoto EM, Safar P, et al. Thiopental amelioration of brain damage after global ischemia in monkeys. Anesthesiology 1978;49(6):390–8.
67. Selman WR, Spetzler RF, Roski RA, et al. Barbiturate coma in focal cerebral ischemia. Relationship of protection to timing of therapy. J Neurosurg 1982;56(5):685–90.
68. Bruno A, Liebeskind D, Hao Q, et al. Diabetes mellitus, acute hyperglycemia, and ischemic stroke. Curr Treat Options Neurol 2010;12(6):492–503.
69. Zhou JW, Wu J, Zhang J, et al. Association of stroke clinical outcomes with coexistence of hyperglycemia and biomarkers of inflammation. J Stroke Cerebrovasc Dis 2015;24(6):1250–5.
70. Rosso C, Pires C, Corvol JC, et al. Hyperglycaemia, insulin therapy and critical penumbral regions for prognosis in acute stroke: further insights from the INSULINFARCT trial. PLoS One 2015;10(3):e0120230.
71. Piironen K, Putaala J, Rosso C, et al. Glucose and acute stroke: evidence for an interlude. Stroke 2012;43(3):898–902.
72. Finfer S, Chittock DR, Su SY, et al. Intensive versus conventional glucose control in critically ill patients. N Engl J Med 2009;360(13):1283–97.
73. Bellolio MF, Gilmore RM, Ganti L. Insulin for glycaemic control in acute ischaemic stroke. Cochrane Database Syst Rev 2014;(1):CD005346.
74. Arnberg F, Grafström J, Lundberg J, et al. Imaging of a clinically relevant stroke model: glucose hypermetabolism revisited. Stroke 2015;46(3):835–42.
75. Capes SE, Hunt D, Malmberg K, et al. Stress hyperglycemia and prognosis of stroke in nondiabetic and diabetic patients: a systematic overview. Stroke 2001;32(10):2426–32.
76. Gill HL, Siracuse JJ, Parrack IK, et al. Complications of the endovascular management of acute ischemic stroke. Vasc Health Risk Manag 2014;10:675–81.
77. Chi OZ, Barsoum S, Rah KH, et al. Local O2 balance in cerebral ischemia-reperfusion improved during pentobarbital compared with isoflurane anesthesia. J Stroke Cerebrovasc Dis 2015;24(6):1196–203.
78. Hassan AE, Chaudhry SA, Zacharatos H, et al. Increased rate of aspiration pneumonia and poor discharge outcome among acute ischemic stroke patients following intubation for endovascular treatment. Neurocrit Care 2012;16(2):246–50.

Neuromonitoring: Martin Smith

Multimodality Neuromonitoring

Matthew A. Kirkman, MBBS, MRCS, MEd, Martin Smith, MBBS, FRCA, FFICM*

KEYWORDS

- Autoregulation • Cerebral perfusion pressure • Cerebrovascular reactivity
- Informatics • Intracranial pressure • Microdialysis • Multimodality neuromonitoring

KEY POINTS

- The clinical neurologic examination is the cornerstone of neuromonitoring, but a complete clinical assessment is not possible in intubated and sedated/anesthetized patients.
- There are several techniques that permit global or regional monitoring of cerebral hemodynamics, oxygenation, metabolism, and electrophysiology, invasively or noninvasively.
- Given the pathophysiological complexity of acute brain injury, a single neuromonitor is unable to detect all instances of cerebral compromise.
- Multimodality neuromonitoring is widely used to individualize patient management after acute brain injury.
- High-quality outcome studies are necessary to demonstrate any outcome effects of multimodal neuromonitoring-guided treatment.

INTRODUCTION

Systemic and central nervous system physiologic monitoring is used to guide the management of patients with neurologic disease in the perioperative and critical care settings. Although the clinical neurologic examination is the cornerstone of neuromonitoring, a complete clinical assessment is not possible in intubated or sedated/anesthetized patients. Several techniques are available for global or regional monitoring of cerebral hemodynamics, oxygenation, metabolism, and electrophysiology to guide patient management.

The pathophysiology of acute brain injury (ABI) is complex, involving changes in cerebral blood flow (CBF), oxygen and glucose delivery and utilization, and electrophysiologic derangements. A single monitoring modality is, therefore, unable to detect

This work is published in collaboration with the Society for Neuroscience in Anesthesiology and Critical Care.

Disclosure Statement: M. Smith is partly funded by the Department of Health National Institute for Health Research Centres funding scheme via the University College London Hospitals/University College London Biomedical Research Centre.

Conflicts of Interest: None.

Neurocritical Care Unit, The National Hospital for Neurology and Neurosurgery, University College London Hospitals, Queen Square, London WC1N 3BG, UK

* Corresponding author.

E-mail address: martin.smith@uclh.nhs.uk

Anesthesiology Clin 34 (2016) 511–523
http://dx.doi.org/10.1016/j.anclin.2016.04.005
1932-2275/16/$ – see front matter © 2016 Elsevier Inc. All rights reserved.
anesthesiology.theclinics.com

all instances of cerebral compromise. Multimodality neuromonitoring is the simultaneous measurement of several variables and provides a more comprehensive picture of the pathophysiology of the injured brain and its response to treatment. It allows an individually tailored approach to the management of patients with ABI in which treatment decisions are guided by monitored changes in pathophysiologic variables rather than generic one-size-fits-all treatment targets. General indications for neuromonitoring are shown in **Box 1**.

This article reviews the neuromonitoring techniques commonly used in perioperative and critical care settings. Some important modalities are covered elsewhere in this edition and are not discussed here. The reader is referred to (see Koht A, Sloan TB: Intraoperative Monitoring: Recent Advances in Motor Evoked Potentials;and Kirkman MA, Smith M: Brain Oxygenation Monitoring, in this issue) for detailed discussions of intraoperative neurophysiological monitoring and brain oxygenation monitoring, respectively.

INTRACRANIAL PRESSURE

The monitoring and management of intracranial pressure (ICP) is the cornerstone of neuromonitoring in patients with traumatic brain injury (TBI) and increasingly used in other brain injury types. In addition to measuring absolute ICP, ICP monitoring permits the calculation of cerebral perfusion pressure (CPP), a therapeutic target in itself; identification and analysis of pathologic ICP waveforms; and derivation of indices of cerebrovascular pressure reactivity.

Intracranial Monitoring Methods

ICP is most commonly measured using an intraventricular catheter or parenchymal microtransducer device.[1] Ventricular catheters measure the pressure of the cerebrospinal fluid (CSF) in the lateral ventricles, which is an assessment of global ICP. This measurement can be achieved using a standard ventricular catheter connected via a fluid-filled system to an external pressure transducer or a catheter incorporating microstrain gauge or fiberoptic technology. Both allow in vivo calibration and therapeutic drainage of CSF.[2] Ventricular catheter insertion can be technically challenging

---

Box 1
Indications for neuromonitoring

- Monitoring the healthy but at-risk brain
  - Intraoperative monitoring during selected procedures, including cardiac and carotid surgery
- Early detection of secondary adverse events after ABI
  - Intracranial hypertension
  - Reduced cerebral perfusion
  - Impaired cerebral glucose delivery/utilization
  - Cerebral hypoxia/ischemia
  - Cellular energy failure
  - Nonconvulsive seizures
- Guiding individualized, patient-specific therapy after ABI
  - Optimization of intracranial and cerebral perfusion pressures
  - Optimization of brain tissue oxygenation
  - Optimization of cerebral glucose delivery/utilization
  - Monitoring cerebral vasospasm after subarachnoid hemorrhage
  - Prognostication

and associated with placement-related hemorrhage and catheter-associated ventriculitis. The risk of ventriculitis increases with time following catheter insertion but can be reduced by the use of antibiotic-impregnated or silver-coated catheters.

Two broad categories of parenchymal microtransducer ICP monitoring systems exist. Solid-state piezoelectric strain gauge devices incorporate pressure-sensitive resistors, which translate pressure-generated changes in resistance to an ICP value. Fiber-optic devices transmit light towards a mirror at the catheter tip, which becomes distorted by changes in ICP. The difference in the intensity of reflected light as the mirror distorts is translated into ICP values. Compared with ventricular catheters, parenchymal ICP measurement devices are easy to insert and have a better safety profile, particularly with regard to hematoma and infection risk.[1] They are usually placed approximately 2 cm into brain parenchyma through a cranial access device or at craniotomy when they can also be sited subdurally. Intraparenchymal devices measure localized pressure but provide equivalent pressure measurements to ventricular catheters in most circumstances. Zero drift, whereby there is a change in baseline ICP readings, is associated with microtransducer systems and can result in measurement error over several days. Furthermore, in vivo recalibration is not possible with parenchymal devices.

Several noninvasive ICP monitoring techniques are available, but most have limitations.[3] A pulsatility index derived from transcranial Doppler (TCD) ultrasonography is an imprecise assessment of ICP compared with invasive measurement alternatives. Optic nerve sheath diameter, measured by ultrasound or computed tomography (CT), is related to ICP and has been used to identify intracranial hypertension. All currently available noninvasive ICP monitoring techniques fail to measure ICP sufficiently accurately for routine clinical use, and most are unable to monitor intracranial dynamics continuously.[4]

Indications for Intracranial Pressure Monitoring

There are no high-quality data confirming an association between ICP-guided management and improved outcomes in any brain injury type. Nevertheless, ICP monitoring is a standard of care after severe TBI in most neuroscience centers.[2] Guidelines from the Brain Trauma Foundation (BTF) published in 2007 contain recommendations for the use of ICP monitoring,[5] and a more recent expert statement provides updated guidance.[6] The key features of these recommendations are summarized in **Box 2**.

Aside from TBI, ICP monitoring provides valuable information to guide the critical care management of aneurysmal subarachnoid hemorrhage (SAH)[7] and intracerebral hemorrhage (ICH).[8] It is also standard in the management of hydrocephalus, including chronic monitoring of normal pressure hydrocephalus, and becoming so in the perioperative management of patients with neoplastic lesions and associated mass effect. However, these indications are not as well defined or well studied as those for TBI.

Intracranial Pressure Monitoring–Guided Treatment

Normal mean ICP is 5 to 10 mm Hg in healthy, resting supine adults. It is generally advised that ICP greater than 20 to 25 mm Hg requires treatment after TBI,[5] but higher and lower thresholds have also been recommended.[2] It is well known that intracranial hypertension is detrimental to outcome, but crucially it is the time spent above a defined ICP threshold as well as absolute ICP values that are important determinants of poor outcome. Changes in the ICP waveform are observed as ICP increases, and waveform analysis has been used to predict the onset of intracranial hypertension.[9]

Box 2
Indications for intracranial pressure monitoring in traumatic brain injury

BTF 2007[5]

- All salvageable patients with severe TBI and an abnormal cranial CT scan
- A normal scan and 2 or more of
  - Age greater than 40 years
  - Unilateral or bilateral motor posturing
  - Systolic blood pressure less than 90 mm Hg

The Milan consensus conference 2014[6]

- ICP should be monitored
  - In comatose patients
    - With cerebral contusions
    - When clinical examination is unreliable
    - When interruption of sedation to check neurologic status is dangerous
  - Following secondary decompressive craniectomy

- ICP monitoring should be considered
  - After evacuation of an acute supratentorial intracranial hematoma in salvageable patients at increased risk of intracranial hypertension, including those with
    - Glasgow coma score motor score of 5 or less
    - Pupillary abnormalities
    - Prolonged/severe hypoxia and/or hypotension
    - Cranial CT findings suggestive of raised ICP
    - Intraoperative brain swelling
    - Or when interruption of sedation to check neurologic status is dangerous
  - In patients with extracranial injuries requiring multiple surgical procedures and/or prolonged analgesia and sedation

---

Evidence that this translates into more timely intervention and improved outcomes is currently lacking.

A meta-analysis incorporating 14 studies of 24,792 patients with severe TBI found that ICP monitoring–guided management of intracranial hypertension was associated with no significant overall mortality benefit compared with treatment without ICP monitoring, although mortality was lower in those who underwent ICP monitoring in studies published after 2012.[10] The only randomized controlled trial of ICP-guided management after TBI (Benchmark Evidence from South American Trials: Treatment of Intracranial Pressure [BEST:TRIP]) found similar 3- and 6-month outcomes in patients in whom treatment was guided by ICP monitoring compared with treatment guided by imaging and clinical examination in the absence of ICP monitoring.[11] Those in the non-ICP monitored group received protocol-specified but empirical treatment on a fixed schedule basis, and the wider applicability of such an approach is questionable given that one of the interventions (mannitol) has been shown to have a more beneficial effect when directed by monitored increases in ICP. In contrast to previous studies,[12] those undergoing ICP monitoring in the BEST:TRIP trial received significantly fewer days of ICP-directed treatment (hyperventilation, hypertonic saline/mannitol, and barbiturates) compared with those in the non-ICP monitored group, although the length of intensive care unit stay was similar. Whether the findings of this study, which was conducted in Bolivia and Ecuador, are applicable to populations with access to superior prehospital care and rehabilitation services remains to be seen. However, this study is important because it reinforces the principle that the evaluation and diagnosis of intracranial hypertension, whether by monitoring ICP or assessment of clinical and imaging variables, is central to the management of patients with severe TBI.

ICP monitoring cannot, and is not designed to, assess the *adequacy* of cerebral perfusion. Several studies confirm that brain hypoxia/ischemia can occur when ICP and CPP are within established thresholds for normality.[13] Moreover, elevated ICP values can arise from both increased CBF (hyperemia) and reduced CBF secondary to cerebral edema, highlighting the nonspecific nature of ICP readings. There is also evidence that multimodality monitoring incorporating brain tissue partial pressure of oxygen ($PtiO_2$) monitoring in addition to ICP can identify cerebral hypoperfusion more reliably than ICP monitoring alone.[14] ICP monitoring is, therefore, best considered as one part of a multimodal neuromonitoring strategy rather than as a monitoring modality in isolation.

Cerebral Perfusion Pressure

CPP is calculated as the difference between mean arterial pressure (MAP) and ICP. Accurate calculation of CPP requires that the zero reference points for both MAP and ICP should be the same, that is, at the level of the brain using the tragus of the ear as the external landmark.[15] Identical reference points for MAP and ICP are particularly important if the head of the bed is elevated, as is routine in the management of ABI. Under such circumstances, measuring arterial blood pressure (ABP) at the level of the heart and ICP at the level of the brain results in an erroneously high calculated CPP; a measured CPP of 60 mm Hg may actually represent a true CPP of less than 45 mm Hg.[15] Such measurement discrepancies are exacerbated in tall patients, with varying elevations of the head of the bed, and different sites of arterial cannulation. Although international management guidelines recommend target values for CPP, the calibration of blood pressure, which directly influences calculated CPP values, is not described. A recent narrative review was unable to determine how MAP was measured in the calculation of CPP in 50% of 32 widely cited studies of CPP-guided management.[16] There have been recent calls for the adoption of international standardization of CPP measurement methods, not only in clinical practice but also in clinical trials.[17]

Cerebral Perfusion Pressure–Guided Treatment

The indications for CPP monitoring are similar to those for ICP, with a predominant application in TBI[15] and emerging indications in other brain injury types.[2]

The recommendations relating to CPP thresholds after TBI have changed over time. The most recent guidelines from the BTF recommend that CPP be maintained between 50 and 70 mm Hg after TBI, with evidence of adverse outcomes with lower or higher values.[5] CPP that is too low risks cerebral hypoperfusion and ischemia, whereas targeting a higher CPP does not guarantee a favorable outcome and is associated with a substantial risk of acute lung injury related to the administration of large fluid volumes and inotropes/vasopressors to increase MAP.[15] Multimodality monitoring incorporating $PtiO_2$ monitoring and autoregulatory indices has been used to identify an optimal CPP value (CPP_{opt}) in an individual patient. Targeting CPP_{opt} rather than a generic CPP threshold minimizes the risks of excessive CPP and associated complications on the one hand and cerebral hypoperfusion and worsening secondary brain injury on the other.[2]

CEREBROVASCULAR REACTIVITY

Cerebrovascular reactivity is a key component of cerebral autoregulation (CA), which can be disturbed or abolished by intracranial pathology as well as by some anesthetic and sedative agents. This reactivity may result in uncoupling of regional CBF and metabolic demand and increase the risk of secondary cerebral ischemia. Standard methods of testing static and dynamic CA are interventional and intermittent with limited applicability in anesthetized or critically ill patients. Methods for the continuous monitoring of cerebrovascular reactivity at the bedside have recently been described.

Pressure Reactivity Index

The pressure reactivity of cerebral vessels determines the ICP response to changes in ABP, with disturbed reactivity implying disturbed pressure autoregulation. In the normal brain, increases in ABP result in cerebral vasoconstriction within 5 to 15 seconds, with an associated reduction in cerebral blood volume (CBV) and ICP. If cerebrovascular reactivity is impaired, CBV and ICP increase passively with ABP, with opposite changes when ABP is reduced. The pressure reactivity index (PRx), calculated as the moving correlation coefficient of consecutive time averaged data points of ICP and ABP over a 4-minute period, can be measured continuously as a marker of autoregulatory status.[18] An inverse correlation between ABP and ICP, indicated by a negative value for PRx, represents normal cerebrovascular reactivity, whereas an increasingly positive PRx defines a continuum of increasingly nonreactive cerebrovascular circulation when changes in ABP and ICP are in phase. After ABI, cerebral vasoreactivity varies with perfusion pressure and optimizes within a narrow range of CPP (CPP_{opt}) specific to an individual patient. Targeting PRx-defined CPP_{opt} allows individualized management of ABP and ICP and minimizes the risks of excessively high or low CPP that can be associated with reliance on a generic CPP threshold.[19]

Other Autoregulatory Indices

Cerebrovascular reactivity can alternatively be assessed with an oxygen reactivity index (ORx), defined as the moving correlation between $PtiO_2$ and CPP. The correlation between ABP and TCD-derived CBF velocity (mean velocity index), and several near-infrared spectroscopy (NIRS)–derived variables (eg, cerebral oximetry index) have also been described.[20] Recently, an innovative technique incorporating ultrasound-tagged NIRS for the measurement of microcirculatory CBF has been used to monitor CA continuously during cardiac surgery.[21]

Autoregulation-Guided Treatment

The assessment of cerebrovascular reactivity, particularly using PRx, has become popular in some centers during the management of TBI and, more recently, after SAH and ICH. Abnormal PRx values, indicating autoregulatory dysfunction, are associated with poor outcome after TBI; in small studies, PRx-guided optimization of CPP has been associated with improved outcomes.[19] A recent systematic review confirmed that monitoring cerebrovascular reactivity, in addition to allowing optimization of CPP, is important in evaluating relationships between CBF, oxygen delivery and demand, and cellular metabolism after TBI.[22]

PRx is considered a global measure of autoregulatory status, whereas ORx represents regional autoregulation because of the focal nature of $PtiO_2$. Thus, findings of deranged ORx but normal PRx after ICH strongly suggests the presence of focal but not global autoregulatory failure.[23]

NIRS-derived measures of cerebrovascular reactivity have also been used to guide brain protection protocols during cardiac surgery. The duration and magnitude of blood pressure less than the lower limit of CA are independently associated with major morbidity and mortality after cardiac surgery according to NIRS-derived data.[24]

CEREBRAL BLOOD FLOW MONITORING

Under normal physiologic conditions cerebral pressure autoregulation maintains CBF constant over a wide range of CPP. CA is often impaired after ABI, and CBF becomes increasingly pressure dependant as autoregulatory failure worsens. Monitoring CBF can, therefore, provide information not only about absolute or relative blood flow but also autoregulatory status.

Kety and Schmidt described the first practical method for measuring CBF in 1945. Their method incorporated the Fick principle and forms the basis of many current CBF measurement techniques. Two bedside methods for the continuous assessment of CBF are available.

Transcranial Doppler Ultrasonography

TCD is an established, noninvasive technique for the real-time assessment of cerebral hemodynamics. Ultrasound waves are used to measure blood flow velocity (FV) through large cerebral vessels from the Doppler shift resulting from red blood cells moving through the field of view. The FV waveform resembles an arterial pressure pulse wave, and waveform analysis permits quantification of peak systolic, end diastolic, and mean FVs. The pulsatility index, which provides an assessment of distal cerebrovascular resistance, can also be measured. The main disadvantages of TCD are its measurement of relative changes as opposed to absolute CBF, and operator dependency. Long-term recordings are limited by the need for accurate and immovable probe fixation; TCD is, therefore, best considered an intermittent monitoring technique.

Indications

TCD has several perioperative indications but is most widely used to monitor changes in cerebral perfusion during carotid endarterectomy. There is good correlation between TCD-monitored variables and electroencephalogram changes of ischemia, and this has been used to guide the need for shunt placement.[25] TCD can also detect emboli as characteristic short-duration, high-intensity chirps; waveform analysis allows differentiation between air and particulate emboli.[26]

TCD is also used in the intensive care management of SAH whereby regular assessments assist in both the diagnosis and management of cerebral vasospasm. Middle cerebral artery (MCA) FV greater than 120 to 140 cm/s or FV increases greater than 50 cm/s/d from baseline indicate developing or established cerebral vasospasm-related delayed cerebral ischemia (DCI). Mean MCA FV thresholds of 100 cm/s and 160 cm/s have been identified as the most accurate thresholds for the detection of angiographic and clinical vasospasm, respectively.[27] Changes in CBF influence FV, and the Lindegaard ratio, which compares ipsilateral MCA and extracranial internal carotid artery FVs, is often preferred to the measurement of FV in a basal cerebral vessel in isolation.[28] Vasospasm is diagnosed by a Lindegaard ratio index greater than 3 and severe spasm by a ratio greater than 6.

Although TCD can detect severe vasospasm with a sensitivity of 97%,[29] it is an operator-dependent tool requiring skilled personnel for interpretation and can only assess a small number of large arteries. Further, in an analysis of 1877 TCD examinations, almost 40% of patients with clinical evidence of DCI never had FVs that exceeded 120 cm/s.[30] Such findings are likely to be related to interindividual variability as well as causes of DCI other than vasospasm, highlighting that treatment decisions in the management of SAH-related DCI should not be based on TCD results alone.

TCD has been used to monitor the integrity of carbon dioxide reactivity as well as pressure autoregulation after ABI and, as noted earlier, to provide a noninvasive but imprecise estimate of ICP. Clinical data on the use of TCD in brain injury types other than TBI and SAH are limited, and there have been concerns about accuracy and reliability.[4]

Thermal Diffusion Flowmetry

Thermal diffusion flowmetry (TDF) is an invasive, continuous, and quantitative monitor of regional CBF. The TDF catheter consists of a thermistor heated to a few degrees greater than the tissue temperature and a second, more proximal, temperature probe. The temperature difference between thermistor and temperature probe is a reflection of heat transfer, which can be translated into a measurement of CBF in milliliters per 100 g/min. A commercial TDF catheter is available, but clinical data using this technology are limited.

CEREBRAL MICRODIALYSIS

Cerebral microdialysis (MD) is a well-established laboratory tool that was introduced as a bedside monitor of brain tissue biochemistry more than 2 decades ago. Because it monitors cellular metabolism as well as substrate supply, MD is able to identify both ischemic and nonischemic causes of cellular energy dysfunction and the ensuing metabolic crisis.[31]

The technical aspects of cerebral MD have been described in detail elsewhere.[32] In brief, a miniature MD catheter is placed into brain tissue and diffusion of molecules across the semipermeable dialysis membrane at its tip allows collection of substances that pass from the brain extracellular fluid (ECF) into the dialysis fluid. This fluid is collected at regular (usually hourly) intervals; the concentrations of glucose, lactate, pyruvate, glycerol, and glutamate can be measured in a semiautomated analyzer at the bedside. Subsequent off-line analysis of the dialysate allows measurement of a myriad of other biomarkers for research purposes.

Interpretation of Cerebral Microdialysis Variables

Each of the biochemical substances measured in the clinical setting is a marker of a particular cellular process associated with glucose metabolism, hypoxia/ischemia, or cellular energy failure.[33] A dramatic increase in cerebral glucose utilization (cerebral hyperglycolysis) may follow ABI, resulting in critical reductions in cerebral glucose levels despite adequate supply. Glucose is the main substrate for brain metabolism, and periods of low cerebral glucose concentration are associated with unfavorable outcomes after TBI.[34]

Glucose is metabolized via glycolysis to pyruvate, which, in the presence of normal oxidative conditions, enters the highly efficient energy-producing tricarboxylic acid (TCA) cycle. Under hypoxic conditions, or if mitochondrial function is compromised, pyruvate is metabolized to lactate outside the TCA cycle, resulting in a lower energy yield. The ECF lactate to pyruvate (LP) ratio is a marker of cellular redox state, and an elevated LP ratio is associated with poor outcome. An increased LP ratio may result from both ischemic and nonischemic causes; it is, therefore, important that absolute lactate and pyruvate concentrations are considered when interpreting the LP ratio. An elevated LP ratio in the presence of low pyruvate (and brain tissue oxygen tension) indicates classic ischemia, whereas an elevated LP ratio in the presence of normal or high pyruvate indicates a nonischemic cause, that is, mitochondrial dysfunction.[35] LP ratio, in combination with ECF glucose levels, therefore, provides useful clinical information about the brain's metabolic state; this ability to assess glucose metabolism is a particular strength of cerebral MD monitoring.[36]

Cerebral MD monitored glutamate is a marker of hypoxia/ischemia and excitotoxicity and glycerol is a marker of hypoxia/ischemia-related cell membrane breakdown.[33]

Because cerebral MD measures change at the cellular level, it may identify cerebral compromise before it is detectable clinically or by other monitored variables.[37]

Indications for Cerebral Microdialysis Monitoring

Cerebral MD should be considered in all patients at risk of developing cerebral hypoxia/ischemia, cellular energy failure, and glucose deprivation.[2] It has been most widely used in the critical care management of TBI and SAH but may also have utility after ICH and acute ischemic stroke. Although the timely detection of impending hypoxia/ischemia would be of significant benefit intraoperatively, a recent systematic review reported limited and low-quality evidence supporting the use of cerebral MD for diagnostic purposes during neurosurgery.[38] Furthermore, the temporal resolution of the only commercially available clinical system (hourly sampling rate) is unlikely to be adequate for intraoperative monitoring. A continuous rapid-sampling cerebral MD technique has been described for research use, but such systems are not currently available for clinical applications.

The heterogeneity of the pathophysiologic changes after ABI means that brain chemistry varies in different regions of the brain. Cerebral MD is a focal technique, so changes in tissue chemistry must be interpreted with knowledge of catheter location. This location can be confirmed by CT visualization of a gold marker at the MD catheter tip. Placement of the catheter in at-risk tissue is generally advocated to facilitate assessment of biochemical changes in the region most susceptible to secondary injury.[39]

Reference Values and Thresholds for Intervention

Absolute normal or abnormal thresholds for monitored brain tissue chemistry are difficult to define based on current data, and a combination of variables has most often been used to relate brain chemistry to outcome. The importance of distinguishing between normal values derived from studies in awake and anesthetized patients undergoing surgery for benign intracranial lesions from those that characterize pathophysiologic disturbance of brain chemistry has recently been emphasized.[39] Further, the trend of variables is as important, or possibly more important, than individual measurements or threshold values.

Cerebral MD monitored glucose, lactate, and LP ratio are now considered more useful in the clinical management of brain-injured patients than glutamate and glycerol.[39] Values that are usually recommended to guide clinical intervention are glucose less than 0.8 mmol/L and LP ratio greater than 40,[34] although a lower LP ratio threshold is recommended by some.[33] When interpreting an elevated LP ratio, lactate concentration greater than 4 mmol/L is generally considered abnormal.[34]

Cerebral hypoglycemia in association with elevated LP ratio is a sign of severe hypoxia/ischemia. If brain glucose is very low (<0.2 mmol/L), a trial of increasing serum glucose (even if within normal limits) should be considered.[39] If the LP ratio indicates ischemia, augmentation of CPP is a therapeutic option, whereas if elevated LP ratio is associated with low brain tissue oxygenation, several interventions that improve oxygen delivery, including judicious increases in CPP or fraction of inspired oxygen or correction of anemia, can be considered. Although the LP ratio has been used to guide CPP optimization after TBI, some studies have found that it may be abnormal despite CPP values that are customarily considered to be adequate.[40] This finding is perhaps unsurprising given the several nonischemic causes of elevated LP ratio, and further highlights the importance of using multimodality physiologic data to guide individualized patient management.

Future Perspectives

Cerebral MD has contributed substantially to our understanding of the pathophysiology of brain injury, but its clinical utility is still debated. Although there is a large body of evidence demonstrating an association between abnormal brain tissue chemistry and clinical outcome after ABI, there are no data to confirm whether cerebral MD-guided therapy can influence outcomes. Future clinical research should focus on assessing the clinical effectiveness of cerebral MD as a component of multimodality monitoring to guide decision-making in acute brain-injured patients and its integration into treatment paradigms in neurocritical care.[39]

NEUROINFORMATICS

Multimodality neuromonitoring produces large and complex datasets. To maximize the clinical effectiveness of monitoring, systems have been developed to analyze and present clinically relevant data in a user-friendly and timely manner at the bedside.[2] Some systems are commercially available, although many have been designed around the needs of individual researchers or institutions.[41] There are several challenges that hinder the integration of data from multiple monitoring modalities, including situations in which one or more monitored variables remain normal in the face of derangements in others and lack of standardization across different monitoring devices, which have often been developed as standalone tools. In the future, incorporation of advanced algorithms is likely to allow automatic recognition and rejection of anticipated and expected fluctuations in data, such as transient increases in ICP associated with suctioning or repositioning of a patient.

There is also interest in the incorporation of computational models of cerebral oxygenation, hemodynamics, and metabolism to interpret complex datasets and provide timely summary outputs that can guide clinical decision-making.[42] Such approaches can also be used to produce patient-specific simulations of clinically important but unmeasured physiologic variables, such as cerebral metabolism. Model-based interpretation of multimodal neuromonitoring data has potential to provide clinicians with information about the underlying processes that are driving the pathophysiologic state of the brain, rather than simply the end points of the injurious processes.

SUMMARY

There is now clear evidence that no single neuromonitor can comprehensively detect all instances of cerebral compromise, and this has driven the development of multimodality neuromonitoring in neurocritical care. The continuous monitoring of multiple physiologic variables, including ICP, CPP, cerebral oxygenation, brain chemistry, and electrophysiology, allows individualized, targeted treatment guided by actual physiologic derangements rather than by generic and often arbitrarily defined thresholds. A multimodal monitoring approach also permits cross-validation between monitored variables and improves confidence in treatment decision-making. However, it remains unclear how a derangement in one physiologic variable in the presence of normal values in others should be managed. Crucial to the widespread implementation of multimodality neuromonitoring is the development of high-quality outcomes studies to determine whether adoption of a multimodal neuromonitoring approach improves outcomes and which modalities are more useful than others in guiding treatment of the acutely injured brain.

REFERENCES

1. Smith M. Monitoring intracranial pressure in traumatic brain injury. Anesth Analg 2008;106:240–8.
2. Le Roux P, Menon DK, Citerio G, et al. Consensus summary statement of the international multidisciplinary consensus conference on multimodality monitoring in neurocritical care: a statement for healthcare professionals from the Neurocritical Care Society and the European Society of Intensive Care Medicine. Intensive Care Med 2014;40:1189–209.
3. Kristiansson H, Nissborg E, Bartek J, et al. Measuring elevated intracranial pressure through noninvasive methods: a review of the literature. J Neurosurg Anesthesiol 2013;25:372–85.
4. Oddo M, Villa F, Citerio G. Brain multimodality monitoring: an update. Curr Opin Crit Care 2012;18:111–8.
5. Brain Trauma Foundation. Guidelines for the management of severe traumatic brain injury. J Neurotrauma 2007;24(Suppl 1):S1–106.
6. Stocchetti N, Picetti E, Berardino M, et al. Clinical applications of intracranial pressure monitoring in traumatic brain injury: report of the Milan consensus conference. Acta Neurochir (Wien) 2014;156:1615–22.
7. Sandsmark DK, Kumar MA, Park S, et al. Multimodal monitoring in subarachnoid hemorrhage. Stroke 2012;43:1440–5.
8. Kirkman MA, Smith M. Supratentorial intracerebral hemorrhage: a review of the underlying pathophysiology and its relevance for multimodality neuromonitoring in neurointensive care. J Neurosurg Anesthesiol 2013;25:228–39.
9. Asgari S, Bergsneider M, Hamilton R, et al. Consistent changes in intracranial pressure waveform morphology induced by acute hypercapnic cerebral vasodilatation. Neurocrit Care 2011;15:55–62.
10. Yuan Q, Wu X, Sun Y, et al. Impact of intracranial pressure monitoring on mortality in patients with traumatic brain injury: a systematic review and meta-analysis. J Neurosurg 2015;122:574–87.
11. Chesnut RM, Temkin N, Carney N, et al. A trial of intracranial-pressure monitoring in traumatic brain injury. N Engl J Med 2012;367:2471–81.
12. Cremer OL, van Dijk GW, van Wensen E, et al. Effect of intracranial pressure monitoring and targeted intensive care on functional outcome after severe head injury. Crit Care Med 2005;33:2207–13.
13. Oddo M, Levine JM, Mackenzie L, et al. Brain hypoxia is associated with short-term outcome after severe traumatic brain injury independently of intracranial hypertension and low cerebral perfusion pressure. Neurosurgery 2011;69:1037–45.
14. Bouzat P, Marques-Vidal P, Zerlauth J-B, et al. Accuracy of brain multimodal monitoring to detect cerebral hypoperfusion after traumatic brain injury. Crit Care Med 2015;43:445–52.
15. Kirkman MA, Smith M. Intracranial pressure monitoring, cerebral perfusion pressure estimation, and ICP/CPP-guided therapy: a standard of care or optional extra after brain injury? Br J Anaesth 2014;112:35–46.
16. Kosty JA, Leroux PD, Levine J, et al. Brief report: a comparison of clinical and research practices in measuring cerebral perfusion pressure: a literature review and practitioner survey. Anesth Analg 2013;117:694–8.
17. Smith M. Cerebral perfusion pressure. Br J Anaesth 2015;115:488–90.

18. Czosnyka M, Pickard JD. Monitoring and interpretation of intracranial pressure. J Neurol Neurosurg Psychiatry 2004;75:813–21.
19. Aries MJH, Czosnyka M, Budohoski KP, et al. Continuous determination of optimal cerebral perfusion pressure in traumatic brain injury. Crit Care Med 2012;40:2456–63.
20. Ono M, Zheng Y, Joshi B, et al. Validation of a stand-alone near-infrared spectroscopy system for monitoring cerebral autoregulation during cardiac surgery. Anesth Analg 2013;116:198–204.
21. Hori D, Hogue CW, Shah A, et al. Cerebral autoregulation monitoring with ultrasound-tagged near-infrared spectroscopy in cardiac surgery patients. Anesth Analg 2015;121:1187–93.
22. Lazaridis C, Andrews CM. Brain tissue oxygenation, lactate-pyruvate ratio, and cerebrovascular pressure reactivity monitoring in severe traumatic brain injury: systematic review and viewpoint. Neurocrit Care 2014;21:345–55.
23. Diedler J, Karpel-Massler G, Sykora M, et al. Autoregulation and brain metabolism in the perihematomal region of spontaneous intracerebral hemorrhage: an observational pilot study. J Neurol Sci 2010;295:16–22.
24. Ono M, Brady K, Easley RB, et al. Duration and magnitude of blood pressure below cerebral autoregulation threshold during cardiopulmonary bypass is associated with major morbidity and operative mortality. J Thorac Cardiovasc Surg 2014;147:483–9.
25. Moritz S, Kasprzak P, Arlt M, et al. Accuracy of cerebral monitoring in detecting cerebral ischemia during carotid endarterectomy: a comparison of transcranial Doppler sonography, near-infrared spectroscopy, stump pressure, and somatosensory evoked potentials. Anesthesiology 2007;107:563–9.
26. Ringelstein EB, Droste DW, Babikian VL, et al. Consensus on microembolus detection by TCD. International consensus group on microembolus detection. Stroke 1998;29:725–9.
27. Mascia L, Fedorko L, terBrugge K, et al. The accuracy of transcranial Doppler to detect vasospasm in patients with aneurysmal subarachnoid hemorrhage. Intensive Care Med 2003;29:1088–94.
28. Lindegaard KF, Nornes H, Bakke SJ, et al. Cerebral vasospasm diagnosis by means of angiography and blood velocity measurements. Acta Neurochir (Wien) 1989;100:12–24.
29. Fontanella M, Valfrè W, Benech F, et al. Vasospasm after SAH due to aneurysm rupture of the anterior circle of Willis: value of TCD monitoring. Neurol Res 2008;30:256–61.
30. Carrera E, Schmidt JM, Oddo M, et al. Transcranial Doppler for predicting delayed cerebral ischemia after subarachnoid hemorrhage. Neurosurgery 2009; 65:316–23 [discussion: 323–4].
31. Vespa P, Bergsneider M, Hattori N, et al. Metabolic crisis without brain ischemia is common after traumatic brain injury: a combined microdialysis and positron emission tomography study. J Cereb Blood Flow Metab 2005;25:763–74.
32. de Lima Oliveira M, Kairalla AC, Fonoff ET, et al. Cerebral microdialysis in traumatic brain injury and subarachnoid hemorrhage: state of the art. Neurocrit Care 2014;21:152–62.
33. Tisdall MM, Smith M. Cerebral microdialysis: research technique or clinical tool. Br J Anaesth 2006;97:18–25.
34. Timofeev I, Carpenter KLH, Nortje J, et al. Cerebral extracellular chemistry and outcome following traumatic brain injury: a microdialysis study of 223 patients. Brain 2011;134:484–94.

35. Larach DB, Kofke WA, Le Roux P. Potential non-hypoxic/ischemic causes of increased cerebral interstitial fluid lactate/pyruvate ratio: a review of available literature. Neurocrit Care 2011;15:609–22.
36. Jalloh I, Carpenter KLH, Helmy A, et al. Glucose metabolism following human traumatic brain injury: methods of assessment and pathophysiological findings. Metab Brain Dis 2015;30:615–32.
37. Belli A, Sen J, Petzold A, et al. Metabolic failure precedes intracranial pressure rises in traumatic brain injury: a microdialysis study. Acta Neurochir (Wien) 2008;150:461–9 [discussion: 470].
38. Bossers SM, de Boer RDH, Boer C, et al. The diagnostic accuracy of brain micro-dialysis during surgery: a qualitative systematic review. Acta Neurochir (Wien) 2013;155:345–53.
39. Hutchinson PJ, Jalloh I, Helmy A, et al. Consensus statement from the 2014 International Microdialysis Forum. Intensive Care Med 2015;41:1517–28.
40. Vespa PM, O'Phelan K, McArthur D, et al. Pericontusional brain tissue exhibits persistent elevation of lactate/pyruvate ratio independent of cerebral perfusion pressure. Crit Care Med 2007;35:1153–60.
41. Sorani MD, Hemphill JC, Morabito D, et al. New approaches to physiological informatics in neurocritical care. Neurocrit Care 2007;7:45–52.
42. Caldwell M, Hapuarachchi T, Highton D, et al. BrainSignals revisited: simplifying a computational model of cerebral physiology. PLoS One 2015;10:e0126695.

Intraoperative Monitoring

Recent Advances in Motor Evoked Potentials

Antoun Koht, MD[a,b,c],*, Tod B. Sloan, MD, MBA, PhD[d]

KEYWORDS

- Intraoperative electrophysiological monitoring • Somatosensory evoked potentials
- Motor evoked potentials • Electroencephalography • Electromyography

KEY POINTS

- Advances in transcranial and direct cortical stimulation have allowed a wider application of motor evoked potentials for mapping and monitoring during procedures on the central nervous system.
- The D-wave amplitude has minimal variability allowing it to be used as a measure of neural injury with mapping and monitoring cortical tumor resection near the motor cortex, and monitoring of surgery on intramedullary spinal cord tumors.
- Stimulation techniques have been developed to measure the proximity of the cortical spinal tract in subcortical and brainstem surgery.
- Conditioning stimuli have been developed to enhance the tcMEP when it is difficult to record.

INTRODUCTION

Advances in electrophysiological monitoring have improved the ability of surgeons to make procedural decisions and reduce complications during surgery and interventional procedures when the central nervous system (CNS) is at risk. Monitoring continues to be done using a multimodality approach using several techniques; each modality provides key aspects for identifying or mapping the location and pathway of critical neural structures or to monitor the progress of procedures so as to reduce the risk of CNS injury. Advances in our understanding of the

This work is published in collaboration with the Society for Neuroscience in Anesthesiology and Critical Care.

Disclosure Statement: The authors have no financial or other conflicts of interest.

[a] Department of Anesthesiology, Feinberg School of Medicine, Northwestern University, 251 East Huron Street, F5-704, Chicago, IL 60611, USA; [b] Department of Neurology, Northwestern University, 251 East Huron Street, F5-704, Chicago, IL 60611, USA; [c] Department of Neurosurgery, Northwestern University, 251 East Huron Street, F5-704, Chicago, IL 60611, USA; [d] Department of Anesthesiology, School of Medicine, University of Colorado, 571 Cambridge Drive, Fairview, TX 75069, USA

* Corresponding author. Department of Anesthesiology, Feinberg School of Medicine, Northwestern University, 251 East Huron Street, F5-704, Chicago, IL 60611.

E-mail address: a-koht@northwestern.edu

electrophysiology of the neural tracts involved have allowed advances in the use of these techniques so that they can evolve to match the needs of complex procedures. Perhaps the most rapidly advancing technique of electrophysiological monitoring is using motor evoked potentials (MEPs).

MOTOR EVOKED POTENTIAL TECHNIQUES

MEPs are produced by stimulation of the motor cortex with recording of responses from the cortical spinal tract (CST) and from the muscles activated by the motor stimulation (**Fig. 1**). The motor cortex can be stimulated by several means. The most

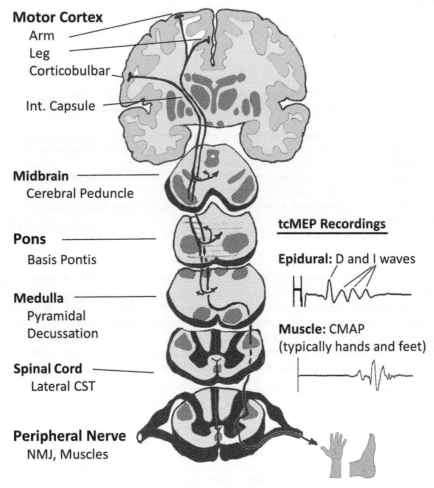

Fig. 1. Stimulation of the motor cortex initiates a traveling wave that descends via corticospinal pathway through the brainstem and through the lateral cortico-spinal pathway to the anterior horn cells in the spinal cord, which results in a muscle contraction through lower motor neurons. The response can be recorded in the epidural space as a D wave and I waves and as a CMAP in the periphery. Stimulation of the corticobulbar motor cortex initiates a traveling wave through the corticobulbar pathway to the brainstem and can be recorded as CMAP in muscles innervated by motor cranial nerves. Int, internal, NMJ, neuromuscular junction.

common technique uses transcranial electrical stimulation with scalp electrodes (tcMEP). Activation of the motor cortex also can be accomplished using the direct application of electrical stimulation to the motor cortex (DCS-MEP).

THE CORTICOSPINAL TRACT

Advances in our understanding of the neuroanatomy of the motor pathways activated by motor cortex stimulation have improved our utilization of MEPs. Transcranial stimulation is thought to activate the motor cortex by direct stimulation of the pyramidal cells and indirectly through synaptic effects in internuncial neurons. This produces a traveling wave in the CST, which consists of a "D" wave (from direct stimulation) and 1 or more "I" waves (from indirect activation). These waves travel in 4% to 5% of the CST, which are fast-conducting fibers.[1] Axons descend from the primary motor and other fronto-parietal cortex in the corona radiata then converge at the internal capsule into a tract that continues down to the spinal cord.

Normally 75% to 90% of the CST crosses the midline at the pyramidal decussation and then descends in the lateral corticospinal tract. The CST is primarily involved with conscious control of skilled movements of the distal extremities and, in particular, facilitation of spinal motor neurons that innervate the distal flexor muscles. Most CST axons coordinate muscular activity through their termination on spinal cord interneurons that eventually reach the anterior horn cell through intermediary synapses. Approximately 2% of the CST neurons directly synapse with anterior horn cells, especially those innervating distal limb muscles, and these represent the key motor function for voluntary muscle activity making them important for monitoring clinical motor function. In addition, these neurons are thought to be more sensitive to ischemic insults than the somatosensory evoked potential pathway (SSEP), reinforcing their value in monitoring.[2]

When the descending volley of D and I waves reach lamina 5 in the spinal cord, they summate to depolarize the anterior horn cells to enable an alpha motor neuron to relay the stimulation down to the neuromuscular junction, which in turn will generate a muscular response.

Several motor tracts besides the CST descend to the spinal cord (including vestibulospinal, reticulospinal, ruprospinal, and tectospinal tracts) and are responsible for the involuntary and automatic control of muscle tone, balance, posture, and locomotion. These do not directly contribute to tcMEPs, but probably influence muscle tcMEPs through background synaptic facilitation and alterations in the excitability of the anterior horn cell.[1,3] Because of varying activity in these pathways, the activation of distal muscles may vary and proximal limb muscles may frequently not be activated by cortical stimulation. As such, the distal limb muscles are preferred for recording MEPs but may vary from stimulation to stimulation.

Advances in our understanding of stimulation have led to recognition of where the motor pathway is actually stimulated. It is important that the stimulation be proximal to the area of neural risk so that the recording is distal and changes represent changes in neural function related to the risk. This is particularly important where the motor cortex is at risk (eg, mapping cortical tumors or identifying motor cortex ischemia). The transcranial method normally stimulates through the anode electrode placed over the motor cortex to be stimulated. However, strong stimulation can result in stimulation of both cortical hemispheres and stimulation of the CST deep within the brain.[1] It has been postulated that subcortical stimulating locations include the internal capsule and brainstem/foramen magnum. The avoidance of subcortical stimulation is paramount for surgery on the cortex, such as tumors and vascular lesions, where the stimulation needs to be in the area at risk or cephalad.

THE CORTICOBULBAR PATHWAY

Stimulation of the lateral motor cortex activates the corticobulbar pathway, which passes along the CST, terminating in the brainstem cranial nerve nuclei. This pathway mediates conscious control over eye, jaw, and face muscles and is involved in swallowing, phonation, and movements of the tongue.

RECORDING D-WAVE RESPONSES

TcMEP can be monitored using the D wave recorded in the epidural space. Because anesthesia impacts neural functioning through changes in synaptic function and no synapses are involved in the D-wave production, the D-wave amplitude usually does not vary more than 10% with static stimuli.[2] As such, its amplitude is proportional to stimulus and reflects the number of recruited fast corticospinal axons.[1] The amplitude gets smaller as the CST descends the spinal cord until insufficient CST is present for a recordable response (at T11–T12).

Advances in D-wave recording have allowed it to be used as a quantitative measure of the CST for procedures involving the motor cortex, subcortical areas, and in spinal cord tumor surgery. Amplitude changes can be used to indicate significant alterations in the motor cortex or CST axons. Of note, the D-wave amplitude is sensitive to the distance of the epidural recording electrode from the CST. Amplitude changes have been seen, without changes in CST function, with spine derotation during the correction of scoliosis when the CST rotates farther from the recording electrode.[4]

In some patients, the D wave may be difficult to record even when muscle tcMEPs can be recorded. This is thought to be due to varying conduction speeds of axons in the CST leading to desynchronization.[3] This happens in approximately 20% of patients with spinal cord pathology including intramedullary spinal cord tumors or after radiation-induced myelopathy. Because the D wave is a product of white matter tracts, motor deficits can occur due to gray matter injury without D-wave changes.[1] Similarly, the D wave is less sensitive to ischemia than muscle tcMEP, as gray matter is more sensitive to ischemia than white matter.[2]

RECORDING MUSCLE RESPONSES

The most commonly recorded tcMEP responses are compound muscle action potentials (CMAP), which can be recorded in muscles innervated by the neural pathway at risk. Advances in our understanding of the neuroanatomy indicate that the distal limb muscles are usually used because (1) the volume of motor gyrus innervating distal limb muscles is much greater than proximal muscles giving rise to better activation, (2) tcMEP preferentially activates the CST to these muscles, and (3) the direct CST innervation of the anterior horn cells for these muscles minimizes anesthetic depression.[5]

The CMAP response is the product of multiple motor units so that the CMAP is usually multipeaked and the amplitude depends on the number of motor units firing. The amplitude is quite variable due to small changes in the anterior horn cell excitability.[1,3]

Because anesthesia has its primary effect at synapses, distal limb muscles innervated by the 2% of CST neurons, which directly connect to the anterior horn cell, are less sensitive to anesthesia than pathways that contain multiple synapses. However, because the production of a muscle response depends on bringing the anterior horn cell to threshold by the summation of D and I waves, loss of I waves from anesthesia effects on the internuncial neurons explains why the currently used high-frequency train technique produces multiple D waves and some I waves, facilitating production of a muscle response.[1,2]

The marked variability of the muscle response gives rise to evolving controversy regarding warning criteria during spinal surgery. The most common alert occurs when a reliably recorded muscle tcMEP is significantly reduced in amplitude (eg, 70%–80%) or can no longer be recorded (ie, "all or none"). Other criteria have been suggested but, until a method to reduce variability emerges, the alert criteria will likely evolve and may be procedure specific.

DIRECT CORTICAL STIMULATION

Advances in direct motor cortex stimulation during craniotomy using grid electrodes or a handheld stimulator (DCS-MEP) have facilitated more focal cortex mapping and monitoring techniques. This technique also helps avoid subcortical activation, which could obscure mapping and monitoring. As a more focal form of stimulation, DCS-MEP usually produces a 1-sided CMAP response seen in the arm or leg contralateral to the side of stimulation. However, its value is limited to monitoring 1 side only and it may require supplementation with transcranial techniques if monitoring of bilateral motor pathways is needed. TcMEP and DCS-MEP do not differ in their capacity to detect an impending lesion of the motor cortex or its efferent pathways.[6]

This stimulation is associated with no or minimal movement of the patient, allowing more frequent monitoring during critical periods when movement would otherwise disrupt the operating surgeon. Other methods used to negate or minimize movement are timing the stimulation with a surgical pause or placing stimulation electrodes closer to the midline to minimize activation of head and neck muscles (unless those muscles are needed for monitoring), and also the controlled use of muscle relaxation discussed later in this article.

MONITORING AND MAPPING THE MOTOR CORTEX

Advances in mapping of the location of the motor cortex can be helpful when pathology, such as epileptogenic tissue, needs to be resected but is near the motor cortex, or when distortion of the cortex obscures the margins of normal and abnormal tissue. This has become important during peri-rolandic surgery, such as surgery of tumors near or in the motor cortex when the procedure cannot be accomplished using an awake craniotomy.[1] Three possible tcMEP methods can be used for monitoring: CMAP, D waves, and I waves. TcMEP assists in mapping the border of the tumor and functioning neural tissue. Monitoring during tumor resection can warn when resection encroaches on motor cortex. Perhaps because motor cortex surgery involves alteration of CST activation, a substantial loss of muscle CMAP amplitude (>75%–80%) has been suggested as the threshold criteria indicating pathway disruption[1]

Quantitative D-wave monitoring can be used because the D wave is generated by the motor cortex pyramidal cells. A 50% loss of D-wave amplitude has been used as warning criteria (30% amplitude loss in a bilaterally generated D wave). Because I waves are produced in the motor cortex, they could also be a logical monitor, but anesthetic suppression limits their usefulness. Their use is being explored using DCS-MEP when they are recorded with a cervical epidural electrode.[3]

MONITORING INTRACRANIAL VASCULAR LESIONS

Because of the ability to detect ischemia, motor monitoring is being used with intracranial aneurysm and arteriovenous malformation procedures. The SSEP is well recognized for this application, but advances in understanding of motor pathway injury from perforating arteries are prompting tcMEP or DCS-MEP monitoring.[7] Studies have shown a clear relationship of tcMEP loss and permanent postoperative neural deficit during aneurysm occlusion of basilar, vertebral, and middle cerebral artery aneurysms.[8] TcMEPs also may be more sensitive and show changes faster than SSEPs for brainstem ischemia caused by perforating artery occlusion during basilar tip aneurysm surgery.[9]

Neurophysiological monitoring is particularly helpful during special procedures with intracranial aneurysms such as temporary clip placement, trapping (occlusion of all arterial supply to isolate an aneurysm), trapping with internal carotid balloon occlusion, and clip reconstruction. In these cases, changes can be unilateral, bilateral, hemispheric, or global. Understanding such procedures by the monitoring team is essential. Changes from temporary clipping are limited to the contralateral side, whereas trapping with multiple temporary clips may result in bilateral changes. Trapping with internal carotid balloon occlusion and temporary clips of ipsilateral anterior, middle, and posterior cerebral arteries may lead to changes on the contralateral hemisphere. In these cases, surgeons may be able to respond to monitoring changes by releasing clips, thereby restoring blood flow and reversing ischemia. Monitoring signals can be used to gauge the safety of these maneuvers. However, during clip reconstruction, the time for completing the reconstruction is key. Changes in tcMEP usually occur approximately 15 to 20 minutes ahead of SSEPs, which can lead to earlier warning and action that may prevent neurologic deficits (Koht et al, Unpublished data, 2016).

MAPPING AND MONITORING SUBCORTICAL PATHOLOGY

Due to advances in our ability to stimulate and record motor pathway responses, there has been a growing interest in subcortical CST mapping (during surgery) to integrate with neuronavigation and diffuser tensor imaging along the corona radiata, internal capsule, and cerebral peduncles. Referred to as subcortical MEP, it involves direct stimulation along the edge of subcortical pathology to assess the proximity of the CST so that the CST is not injured during lesion resection. Using a handheld sterile stimulator, the distance from the CST is measured as 1.0 to 1.5 mA of stimulation for each millimeter of distance.[10] This dynamic mapping avoids the effect of tissue distortion as brain shift occurs making neuronavigation less accurate. Using this proximity mapping, a 2-mA to 5-mA threshold has been used to halt surgery. This allows a "safe" corridor with sufficient distance from the CST to avoid iatrogenic CST damage.[10]

MAPPING AND MONITORING THE BRAINSTEM

Monitoring of procedures on brainstem structures usually involves recording the muscle responses to stimulation of the brainstem nuclei or cranial nerves serving those monitored muscles. This technique, however, requires the participation of the surgeon and leaves gaps in the monitoring when the surgeon is not stimulating (unless they are using electrified operating instruments connected to a source of stimulation). Advances in evoked responses produced by transcranial stimulation of the corticobulbar motor cortex (Co-MEPs) have allowed continual assessment of the cranial nerve motor pathway. This also includes nerve portions that are proximal to the surgical site, which might be at risk with intra-axial tumors or with large extra-axial tumors.[11]

At present, methods have been described for monitoring of the vagus/recurrent laryngeal nerve (cranial nerve [CN] X) and facial nerve (CN VII). Co-MEPs in the larynx (CN X) are best done using hook-wire electrodes in the vocalis muscle and have been successfully used during skull base surgery, carotid endarterectomy, anterior spine surgery, thyroid resection, neck tumors, and craniotomies in the motor area.[12] Unfortunately, larger recording electrodes (such as those of specialized endotracheal tubes) can detect far field activity of other neck muscles also activated by cortical stimulation and potentially miss a loss of response in the cricothyroid muscle.

Techniques for Co-MEP monitoring of the facial nerve have also been reported in surgeries including those of cranial base, brainstem, posterior fossa, and for large cerebello pontine angle (CPA) tumors where proximal access to CN VII is not permitted.[11,13] Transcranial stimulation could also activate the facial nerve extracranially bypassing the corticobulbar tracts. To help distinguish facial muscle activity from this direct stimulation, the muscle response from single pulse transcranial stimulation can be compared with the usual multipulse technique at the same stimulation intensity. The single pulse may activate the muscle directly but will be blocked in the corticobulbar pathway by the effects of anesthesia.[13,14]

In addition to monitoring specific cranial nuclei, general monitoring of the brainstem also helps avoid injury. Monitoring of the SSEP and auditory brainstem response provides vigilance for only approximately 20% of the cross-section of the brainstem. Adding monitoring of the tcMEP increases the brainstem tissue monitored and is thought to be of value to reduce motor tract injuries, particularly with the removal of lesions in the midbrain near the cerebral peduncle (floor of the fourth ventricle).[14] In these surgeries, the CST also can be mapped using a handheld stimulating probe with recording of the D wave using lower cervical/upper thoracic epidural electrodes (similar to subcortical techniques mentioned previously).

MONITORING SPINAL SURGERY

It has been known for some time that the risk of paralysis during the correction of scoliosis is reduced using monitoring.[15] This assessment was conducted when monitoring was largely conducted with SSEP, which would occasionally fail to warn of a motor deficit. Since the introduction of tcMEP monitoring, an evidence-based review reinforced the value of monitoring in spine surgery.[16] Due to advances in tcMEP, this technique has been embraced by spine and neurologic surgeons as a method to help improve outcome and reduce risk in a wide variety of surgeries for correction of axial skeletal deformities.

The muscle response of the tcMEP is more sensitive to CST ischemia because it involves synapses in the spinal gray matter.[17] The white matter of the SSEP and CST (including the D wave) are less sensitive. Given this sensitivity to ischemia and the relative greater importance of motor function, tcMEP monitoring is being used to monitor the vascular supply of the spinal cord in situations such as surgery on thoraco-abdominal aneurysms or spinal arteriovenous malformations. In the latter, the importance of a feeding artery can be determined by test clamping or by the effect on the tcMEP and SSEP by the injection of sodium amytal (which blocks synaptic transmission of the tcMEP) or lidocaine (which blocks axonal conduction of the SSEP and tcMEP).[18] Experience with tcMEP monitoring during spine surgery has changed our concept of the safe acceptable blood pressure, as alterations in tcMEP responses have been seen at blood pressures previously thought to be acceptable.

MONITORING INTRAMEDULLARY SPINAL CORD TUMORS

Advances in tcMEP monitoring have resulted in its role in improving long-term motor function with intramedullary spinal cord tumor resection.[19] For monitoring, the D wave and muscle responses of the tcMEP are used (the SSEP is frequently lost during a midline myelotomy).[1] Because many of these tumors are benign and in the cervical region, the focus is on removing as much tumor as possible without causing undue motor tract injury. The D wave is used as a semiquantitative measure of the CST.[2] As long as the D-wave amplitude remains higher than 50%, motor function will be preserved, even if the tcMEP muscle response is lost (in which case a temporary deficit occurs).[2,19] In the absence of a D wave, muscle tcMEPs are monitored with a limit of 50% amplitude drop.[2]

Methods are currently being developed to localize the CST within the spinal cord using a handheld stimulation probe. Current study is focused on "collision" studies in which stimulation of the spinal cord using a probe blocks conduction in the CST such that a distal D wave and muscle CMAP is prevented when the probe is near the CST.[20,21]

DEVELOPMENT OF ENHANCING MOTOR STIMULATION TECHNIQUES

Advances in our understanding of the effects of anesthesia and the development of total intravenous anesthetic techniques have markedly increased the ability to routinely record tcMEP during surgery. In particular, refinements in propofol and opioid infusions with supplementation with ketamine and lidocaine have improved success of tcMEP recording and patient management. However, as surgical procedures have expanded into more complex procedures, more patients present with preexisting neurologic deficits in which tcMEPs are difficult to acquire. These patients include adults with diabetes, spinal cord or nerve root injury, vascular disease, cerebral palsy, brain injury, and muscle or axonal conduction impairment. This group also includes individuals in whom anesthesia management is difficult to conduct with total intravenous anesthesia (TIVA) and children who have an immature CNS.[6]

Advances in tcMEP stimulation techniques have been helpful in acquiring responses in these patients. These techniques use stimulation paradigms using enhancement by using a preconditioning stimulus before the test stimulus. Two types of conditioning stimulation have been explored. The first type is referred to as homonymous stimulation because both conditioning and test stimuli are applied at the same site. The most common technique uses a preconditioning tcMEP stimulation train similar to the test train (referred to as double-train tcMEP).[1] The effect is thought to build up alpha motor neuron excitability and recruit additional neuron pools. The time between the conditioning and test train is referred to as the intertrain interval (ITI). Facilitation appears to be most effective when the muscle responses would otherwise be small or absent.[22] The best facilitation occurs with an ITI of 10 to 35 ms with an inhibitory effect at longer ITI, and then a second window of enhancement at still longer ITI (100–1000 ms).

Simply conducting a series of tcMEP test stimuli can act as a conditioning stimulus if the rate of testing has a favorable ITI (1–10 Hz). In general, the faster the tcMEP monitoring rate in this range, the higher the tcMEP muscle amplitude, but a 2-Hz rate has been suggested to minimize possible cortical injury.[2]

The second type of preconditioning stimuli is referred to as heteronymous conditioning because the stimuli are delivered at different sites. The effect appears to be through central mechanisms for some stimuli, as enhancement can be reduced by lorazepam.[23] This suggests some techniques of enhancement may be affected by anesthesia choice.

Several types of heteronymous conditioning stimuli have been used:

1. Stimulation of the plantar region of the foot or palmar region of the hand with pulse trains.[24]
2. Peripheral sensory nerve stimulation (thought to produce lower motor neuron facilitation). This effect is segmental and lateralized, so that one might be able to focally enhance tcMEPs of the arms or legs.[3]
3. High-frequency foot sole stimulation 50 to 100 ms before tcMEP, producing a subliminal withdrawal reflex.[25] This facilitates tibialis anterior responses so much that they can be evoked with a single transcranial pulse.
4. Peripheral sensory stimulus in the receptive field of the withdrawal reflex of the anterior tibialis muscle.[26]
5. Tetanic stimulation, as discussed later in this article.

MONITORING DURING PARTIAL NEUROMUSCULAR BLOCKADE

Because of patient movement and difficulties in providing anesthesia for the tcMEP technique, there is a growing interest in the use of partial muscle relaxation (pNMB). Ample experience with pNMB has demonstrated that this can be used when the patient otherwise has robust tcMEP muscle responses and the residual muscle response after pNMB is adequate as long as a steady state degree of relaxation is maintained.[27] However, in patients with poor responses, this may prevent monitoring and has led to the usual recommendation to avoid neuromuscular blockade during monitoring. When pNMB must be used, tetanic stimulation to motor nerves has been studied as a means of enhancing the tcMEP muscle responses similar to the effect of tetanic stimulation on traditional assessment of neuromuscular blockade. Studies show the percentage of patients with recordable responses and muscle amplitudes are both increased following tetanic stimulation.[28]

Interestingly, tetanic stimulation produces tcMEP amplitude enhancement in muscles in addition to those innervated by the stimulated nerve, suggesting central enhancement.[28] Unfortunately, this central enhancement by tetany does not occur in the presence of sensory deficits or motor dysfunction.[28]

SUMMARY

The techniques of electrophysiological mapping and monitoring continue to evolve to match the need of procedures in which the CNS is at risk. Like other tools in the surgeon's and interventional physician's tool box, its role in patient care will likely continue to increase. Advances in MEPs have expanded their use for mapping and monitoring during procedures.

REFERENCES

1. Macdonald DB. Intraoperative motor evoked potential monitoring: overview and update. J Clin Monit Comput 2006;20(5):347–77.
2. Deletis V, Sala F. Intraoperative neurophysiological monitoring of the spinal cord during spinal cord and spine surgery: a review focus on the corticospinal tracts. Clin Neurophysiol 2008;119(2):248–64.
3. Macdonald DB, Skinner S, Shils J, et al. Intraoperative motor evoked potential monitoring—a position statement by the American Society of Neurophysiological Monitoring. Clin Neurophysiol 2013;124(12):2291–316.

4. Ulkatan S, Neuwirth M, Bitan F, et al. Monitoring of scoliosis surgery with epidurally recorded motor evoked potentials (D wave) revealed false results. Clin Neurophysiol 2006;117(9):2093–101.
5. Taniguchi M, Cedzich C, Schramm J. Modification of cortical stimulation for motor evoked potentials under general anesthesia: technical description. Neurosurgery 1993;32(2):219–26.
6. Sala F, Manganotti P, Grossauer S, et al. Intraoperative neurophysiology of the motor system in children: a tailored approach. Childs Nerv Syst 2010;26(4): 473–90.
7. Guo L, Gelb AW. The use of motor evoked potential monitoring during cerebral aneurysm surgery to predict pure motor deficits due to subcortical ischemia. Clin Neurophysiol 2011;122(4):648–55.
8. Szelenyi A, Langer D, Beck J, et al. Transcranial and direct cortical stimulation for motor evoked potential monitoring in intracerebral aneurysm surgery. Neurophysiol Clin 2007;37(6):391–8.
9. Quinones-Hinojosa A, Alam M, Lyon R, et al. Transcranial motor evoked potentials during basilar artery aneurysm surgery: technique application for 30 consecutive patients. Neurosurgery 2004;54(4):916–24.
10. Landazuri P, Eccher M. Simultaneous direct cortical motor evoked potential monitoring and subcortical mapping for motor pathway preservation during brain tumor surgery: is it useful? J Clin Neurophysiol 2013;30(6):623–5.
11. Dong CC, Macdonald DB, Akagami R, et al. Intraoperative facial motor evoked potential monitoring with transcranial electrical stimulation during skull base surgery. Clin Neurophysiol 2005;116(3):588–96.
12. Deletis V, Fernandez-Conejero I, Ulkatan S, et al. Methodology for intraoperatively eliciting motor evoked potentials in the vocal muscles by electrical stimulation of the corticobulbar tract. Clin Neurophysiol 2009;120(2):336–41.
13. Acioly MA, Liebsch M, Carvalho CH, et al. Transcranial electrocortical stimulation to monitor the facial nerve motor function during cerebellopontine angle surgery. Neurosurgery 2010;66(6 Suppl Operative):354–61 [discussion: 62].
14. Morota N, Ihara S, Deletis V. Intraoperative neurophysiology for surgery in and around the brainstem: role of brainstem mapping and corticobulbar tract motor-evoked potential monitoring. Childs Nerv Syst 2010;26(4):513–21.
15. Nuwer MR, Dawson EG, Carlson LG, et al. Somatosensory evoked potential spinal cord monitoring reduces neurologic deficits after scoliosis surgery: results of a large multicenter survey. Electroencephalogr Clin Neurophysiol 1995;96(1): 6–11.
16. Nuwer MR, Emerson RG, Galloway G, et al. Evidence-based guideline update: intraoperative spinal monitoring with somatosensory and transcranial electrical motor evoked potentials: report of the Therapeutics and Technology Assessment Subcommittee of the American Academy of Neurology and the American Clinical Neurophysiology Society. Neurology 2012;78(8):585–9.
17. Sloan TB, Jameson LC. Electrophysiologic monitoring during surgery to repair the thoraco-abdominal aorta. J Clin Neurophysiol 2007;24(4):316–27.
18. Doppman JL, Girton M, Oldfield EH. Spinal Wada test. Radiology 1986;161(2): 319–21.
19. Sala F, Palandri G, Basso E, et al. Motor evoked potential monitoring improves outcome after surgery for intramedullary spinal cord tumors: a historical control study. Neurosurgery 2006;58(6):1129–43 [discussion: 43].
20. Deletis V, Bueno De Camargo A. Interventional neurophysiological mapping during spinal cord procedures. Stereotact Funct Neurosurg 2001;77(1–4):25–8.

21. Deletis V. Intraoperative neurophysiology of the corticospinal tract of the spinal cord. Suppl Clin Neurophysiol 2006;59:107–12.
22. Journee HL, Polak HE, de Kleuver M, et al. Improved neuromonitoring during spinal surgery using double-train transcranial electrical stimulation. Med Biol Eng Comput 2004;42(1):110–3.
23. Kaelin-Lang A, Luft AR, Sawaki L, et al. Modulation of human corticomotor excitability by somatosensory input. J Physiol 2002;540:623–33.
24. Journee HL, Hoving EW, Mooij JJA. Stimulation threshold-age relationship and improvement of muscle potentials by preconditioning transcranial stimulation in young children. Clin Neurophysiol 2006;117:235–9.
25. Cruccu G, Inghilleri M, Berardelli A, et al. Cortical mechanisms mediating the inhibitory period after magnetic stimulation of the facial motor area. Muscle Nerve 1997;20(4):418–24.
26. Andersson G, Ohlin A. Spatial facilitation of motor evoked responses in monitoring during spinal surgery. Clin Neurophysiol 1999;110(4):720–4.
27. Sloan TB. Muscle relaxant use during intraoperative neurophysiologic monitoring. J Clin Monit Comput 2013;27(1):35–46.
28. Hayashi H, Kawaguchi M, Yamamoto Y, et al. Evaluation of reliability of post-tetanic motor-evoked potential monitoring during spinal surgery under general anesthesia. Stereotact Funct Neurosurg 2008;33(26):E994–1000.

Brain Oxygenation Monitoring

Matthew A. Kirkman, MBBS, MRCS, MEd, Martin Smith, MBBS, FRCA, FFICM*

KEYWORDS

- Cerebral oxygenation • Cerebral perfusion • Brain monitoring
- Brain tissue oxygen tension • Jugular venous saturation
- Near-infrared spectroscopy

KEY POINTS

- The maintenance of adequate cerebral oxygenation is a key goal in the management of patients with acute brain injury (ABI) and in certain perioperative settings.
- A mismatch between cerebral oxygen supply and demand can lead to cerebral hypoxia/ischemia and deleterious outcomes; cerebral oxygenation monitoring is, therefore, an important aspect of multimodality neuromonitoring.
- There is abundant evidence of an association between low cerebral oxygenation and outcomes, but limited evidence that increasing cerebral oxygenation improves outcome.
- Advances in cerebral oxygenation monitoring will be driven by improved technology and randomized studies proving the utility of different monitors.

INTRODUCTION

Maintenance of cerebral oxygen supply sufficient to meet metabolic demand is a key goal in the management of patients with ABI and in perioperative settings. A mismatch between oxygen supply and demand can lead to cerebral hypoxia/ischemia and deleterious outcomes, with time-critical windows to prevent or minimize permanent ischemic neurologic injury. The clinical manifestations of cerebral hypoxia/ischemia may remain occult in unconscious or sedated/anesthetized patients, and brain monitoring is required to detect impaired cerebral oxygenation in such circumstances.

Cerebral oxygenation monitoring assesses the balance between cerebral oxygen delivery and utilization, and, therefore, the adequacy of cerebral perfusion and oxygen

This work is published in collaboration with the Society for Neuroscience in Anesthesiology and Critical Care.
Disclosure Statement: M. Smith is part funded by the Department of Health National Institute for Health Research Centres funding scheme via the University College London Hospitals/University College London Biomedical Research Centre.
Conflicts of Interest: None.
Neurocritical Care Unit, The National Hospital for Neurology and Neurosurgery, University College London Hospitals, Queen Square, London WC1N 3BG, UK
* Corresponding author.
E-mail address: martin.smith@uclh.nhs.uk

Anesthesiology Clin 34 (2016) 537–556
http://dx.doi.org/10.1016/j.anclin.2016.04.007
1932-2275/16/$ – see front matter © 2016 Elsevier Inc. All rights reserved.

anesthesiology.theclinics.com

delivery. It can be used to guide treatment to prevent or minimize cerebral hypoxia/ischemia, and is established as an important component of multimodality neuromonitoring in both perioperative and ICU settings.

This article describes the different methods of bedside cerebral oxygenation monitoring, the indications and evidence base for their use, and limitations and future perspectives.

METHODS OF MONITORING CEREBRAL OXYGENATION

There exist several imaging and bedside methods of monitoring global and regional cerebral oxygenation, invasively and noninvasively (**Table 1**). Different monitors describe different physiologic variables and, for this reason, they are not interchangeable.

Imaging Techniques

In addition to providing structural information, several imaging techniques are able to evaluate cerebral hemodynamics and metabolism over multiple regions of interest. Imaging provides only a snapshot of cerebral physiology at a particular moment in time and may miss clinically significant episodes of cerebral hypoxia/ischemia, so continuous, bedside monitoring modalities are preferred during clinical management. Readers are referred elsewhere for a detailed description of the role of imaging after ABI.[1]

Jugular Venous Oxygen Saturation Monitoring

Jugular venous oxygen saturation monitoring ($Sjvo_2$) was the first bedside monitor of cerebral oxygenation, but its use is being superseded by other monitoring tools.

Table 1
Bedside monitors of cerebral oxygenation

	Advantages	Disadvantages
$Sjvo_2$	• Real time • Global trend monitor	• Invasive insertion procedure with risk of hematoma, carotid puncture, and vein thrombosis during prolonged monitoring • Insensitive to regional ischemia • Assumes stable $CMRo_2$ to infer CBF changes
$Ptio_2$	• Focal monitor permitting selective monitoring of critically perfused tissue • Real time • The most effective bedside method of detecting cerebral ischemia • Relatively safe with low hematoma rate (<2%, usually small and clinically insignificant) • No reported infections	• Focal monitor – the position of the probe is crucial • May miss important pathology distant from the monitored site • Invasive • Small degree of zero and sensitivity drift • One-hour run-in period required and thus critical early hypoxic/ischemic episodes may go undetected • Technical complication rates (dislocation or drift) may reach 13.6%
NIRS	• Real time • High spatial and temporal resolution • Noninvasive • Assessment of several regions of interest simultaneously	• Extracerebral circulation may contaminate cerebral oxygenation measurements • Lack of standardization between commercial devices • Thresholds for cerebral hypoxia/ischemia undetermined • Current devices only monitor relative changes in oxygenation

Svjo$_2$ can be measured through intermittent sampling of blood from a catheter with its tip sited in the jugular venous bulb, or continuously using a fiberoptic catheter. Svjo$_2$ represents a global measure of cerebral oxygenation and provides a nonquantitative estimate of the adequacy of cerebral perfusion based on the simple tenet that increased cerebral oxygen demand in the face of inadequate supply increases the proportion of oxygen extracted from hemoglobin and thus reduces the oxygen saturation of blood draining from the brain.[2] The range of normal Svjo$_2$ values is 55% to 75%, and interpretation of changes is straightforward. Low Svjo$_2$ values may indicate cerebral hypoperfusion secondary to decreased cerebral perfusion pressure (CPP) or hypocapnea, or increased oxygen demand that is not matched by increased supply, whereas high values may indicate relative hyperemia or arteriovenous shunting. The arterial to jugular venous oxygen content concentration difference, and other derived variables, have been studied extensively as an assessment of CBF.[3]

Sjvo$_2$ monitoring has been used during cardiac surgery and craniotomy, although its primary role is in the neuro-ICU, where it has been used to detect impaired cerebral perfusion after traumatic brain injury (TBI) and subarachnoid hemorrhage (SAH), to optimize CPP and, historically, to guide therapeutic hyperventilation. No interventional trials, however, have confirmed a direct benefit of Sjvo$_2$-directed therapy on outcome, and there are several limitations associated with its use (see **Table 1**). It may miss critical regional ischemia because it is a global, flow-weighted measure. Furthermore, high Sjvo$_2$ values are not necessarily reassuring because they may be associated with pathologic arteriovenous shunting, and brain death.[2]

Brain Tissue Oxygen Tension Monitoring

Recent years have seen an increasing trend toward the direct measurement of brain tissue partial pressure of oxygen (Ptio$_2$), particularly in patients in whom intracranial pressure (ICP) monitoring is indicated. Ptio$_2$ monitoring has the most robust evidence base of all cerebral oxygenation monitors and is now considered the gold standard for monitoring cerebral oxygenation at the bedside.[4] It has contributed significantly to the understanding of the pathophysiology of ABI and emphasized the importance of multi-modality neuromonitoring.[5]

Evidence from studies of patients with TBI demonstrate that cerebral hypoxia/ischemia can occur when ICP and CPP are within established thresholds for normality.[6] As well as highlighting that ICP and CPP do not directly assess the adequacy of cerebral perfusion, these data suggest that reliance on a single monitoring modality is insufficient to detect cerebral compromise.[7] Ptio$_2$ monitoring has also challenged the role of some components of triple-H therapy in the treatment of SAH,[8] such that induced hypertension alone is now preferred in patients with suspected delayed cerebral ischemia (DCI).

Technical aspects

Ptio$_2$ catheters are similar in size to intraparenchymal ICP monitors and are placed in subcortical white matter through single or multiple lumen bolts, via a burr hole, or at craniotomy. Ptio$_2$ readings are unreliable in the first hour after insertion, and a run-in period is essential. This limits intraoperative applications unless the monitor is already in situ. Correct functioning of the probe is confirmed prior to commencing monitoring through an oxygen challenge, which should be repeated on a daily basis thereafter. A normal probe response is an increase of 200% or more from baseline Ptio$_2$ after an increase in fraction of inspired oxygen (Fio$_2$) to 1.0 for approximately 20 minutes, although impaired pulmonary function can affect responsiveness.

Ptio$_2$ is a focal measure; the region of interest interrogated by a Ptio$_2$ probe is approximately 17 mm^2. Probe placement is, therefore, crucial, and location in at-risk and viable brain tissue is considered optimal by many (**Fig. 1**). Thus, in patients with focal lesions, such as intracerebral hemorrhage (ICH) or traumatic contusions, a perilesional location is favored, whereas in aneurysmal SAH, probe placement in appropriate vascular territories is advised. Such precise placement can be technically challenging or impossible and risks inadvertent intralesional placement, which does not yield useful information. There is, therefore, an argument for routine Ptio$_2$ probe placement in normal-appearing brain, typically in the nondominant frontal lobe,

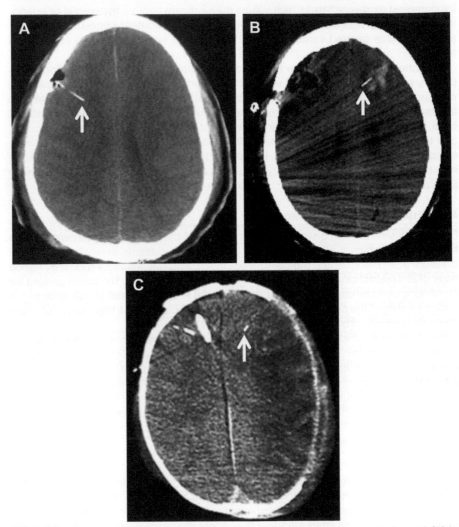

Fig. 1. Axial CT scan of the head demonstrating positioning of a Ptio$_2$ probe (*arrow*) (*A*) in normal-appearing white matter of the right frontal lobe to measure global cerebral oxygenation; (*B*) suboptimally within a contusion, which does not yield useful information; and (*C*) adjacent to a penumbral region to monitor cerebral oxygenation in at-risk tissue. (*From* Le Roux PD, Oddo M. Parenchymal brain oxygen monitoring in the neurocritical care unit. Neurosurg Clin N Am 2013;24(3):431; with permission.)

when it effectively acts as a global measure of cerebral oxygenation and can guide maintenance of normal physiologic function in uninjured brain. This is the preferred site in cases of diffuse brain injury. Satisfactory probe location must always be confirmed with a nonenhanced cranial CT scan to allow appropriate interpretation of $Ptio_2$ readings. One major caveat of $Ptio_2$ monitoring is that heterogeneity of brain oxygenation, even in undamaged areas of brain, is well recognized.[9]

Indications for brain tissue oxygen tension monitoring

$Ptio_2$ monitoring has been used in both ICU and perioperative settings (**Table 2**), but its primary role, and the one for which there is most evidence, is in the intensive care management of severe TBI.[6,10] Recent guidelines from the Neurocritical Care Society recommend that $Ptio_2$ monitoring can be used to titrate individual targets for CPP, arterial partial pressure of carbon dioxide ($Paco_2$), arterial partial pressure of oxygen (Pao_2), and hemoglobin concentration, and to manage intracranial hypertension in combination with ICP monitoring.[11] The Brain Trauma Foundation recommends monitoring and managing $Ptio_2$ as a complement to ICP/CPP-guided treatment in patients with severe TBI.[12] It has also been used in poor-grade aneurysmal SAH[13] and in ICH[14] and has recently been recommended by the Neurocritical Care Society as a means of detecting DCI in sedated or poor-grade SAH patients.[15]

There have been several published reports of intraoperative $Ptio_2$ monitoring but no established role for this indication (see **Table 2**). A patient with an indwelling $Ptio_2$ monitor already in place, however, should have monitoring continued during an operative procedure.

Normal brain tissue oxygen tension values and thresholds for treatment

$Ptio_2$ is a complex and dynamic variable representing the interaction between cerebral oxygen delivery and demand[4] as well as tissue oxygen diffusion gradients.[16] Both cerebral and systemic factors influence $Ptio_2$ values (**Box 1**), and $Ptio_2$ is best considered a biomarker of cellular function as opposed to a simple monitor of hypoxia/ischemia. This makes it an appropriate therapeutic target.

Normal brain $Ptio_2$ is reported to lie between 20 mm Hg and 35 mm Hg (2.66 kPa and 4.66 kPa). PET studies of patients with TBI suggest that the ischemic threshold is less than 14 mm Hg (<1.86 kPa)[17] and clinical studies indicate that $Ptio_2$ below 10 mm Hg (<1.33 kPa) should be considered an indicator of severe brain hypoxia.[9] The threshold for treatment of low $Ptio_2$ remains, however, undecided, with recommendations for initiation of treatment varying from $Ptio_2$ less than or equal to 20 mm Hg ($\leq$2.66 kPa) to less than or equal to 15 mm Hg ($\leq$2 kPa). It is important to appreciate that these thresholds have been determined through patient outcomes in small studies rather than pathophysiologic evidence of ischemic damage at the cellular level.

Crucial to the interpretation of $Ptio_2$ in the clinical setting is the severity, duration, and chronologic trend of cerebral hypoxia and not absolute $Ptio_2$ values in isolation, because it is the overall burden of hypoxia/ischemia that is the key determinant of outcome.[6,9] Many uncertainties remain about when and how to treat reduced $Ptio_2$, and this should be the focus of future research efforts. In particular, the efficacy and safety of increasing Fio_2 to normalize $Ptio_2$ are uncertain given the (low-quality) evidence of a relationship between arterial hyperoxia and increased mortality after ABI.[18] It also remains to be determined whether $Ptio_2$ values above a certain threshold ensure adequate cerebral oxygenation or whether or how a mildly elevated ICP should be managed in the presence of normal $Ptio_2$ values (discussed later).

Table 2
The main indications for brain tissue oxygen tension monitoring and near-infrared spectroscopy in the perioperative and intensive care settings, based on clinical data

Modality	Indications	Evidence	Highest Quality Evidence
Ptio$_2$	*ICU*		
	Severe TBI	Low Ptio$_2$ is associated with worse mortality,[47] lower GOS,[6,9] and increased neuropsychological deficits.[48]	Prospective observational
		Treatment of low Ptio$_2$ may improve outcomes.[10,20]	Randomized controlled trial
		Ptio$_2$ can help define individual CPP thresholds.[49]	Prospective observational
		Response to Ptio$_2$-guided therapy is associated with reduced mortality.[50]	Retrospective analysis of prospective observational data
	Poor-grade SAH	Low Ptio$_2$ values are associated with increased mortality, but the relationship with morbidity is less clear.[51]	Prospective observational
		Ptio$_2$-derived ORx autoregulation assessment can predict the risk of DCI[13] and unfavorable outcome.[52]	Prospective observational
		Response to Ptio$_2$-guided therapy is associated with improved long-term functional outcomes.[19]	Retrospective analysis of prospective observational data
	ICH	Ptio$_2$ monitoring may help identify optimal CPP targets.[53]	Retrospective
		Reduced perihematomal Ptio$_2$ values are associated with poor outcome.[53]	Retrospective
	AIS	Ptio$_2$-derived CPP-ORx may predict the development of malignant MCA infarction.[54]	Prospective observational
	Perioperative		
	Cerebral angiography	Low Ptio$_2$ values are correlated with severe intracranial angiographic arterial caliber reduction in patients with poor-grade SAH.[55]	Retrospective
	Aneurysm surgery	Ptio$_2$ threshold of 15 mm Hg is found a sensitive indicator of the likelihood of developing procedure-related ischemia.[56]	Prospective observational
	AVM surgery	Correlation is observed between reduced Ptio$_2$ values and development of a periprobe ischemic infarction.[57]	Case report

NIRS			
ICU			
	TBI	There is an association between increasing length of time with rSco$_2$ ≤60% and mortality, intracranial hypertension, and compromised CPP.[58]	Prospective observational
	SAH	Time-resolved NIRS was able to predict angiographic-proved vasospasm with 100% sensitivity and 85.7% specificity and confirm vasospasm when TCD was not diagnostic.[59]	Prospective observational
	AIS	May help predict cerebral edema in patients with complete MCA infarction.[33]	Prospective observational
		rSco$_2$ predicts poor outcome during endovascular therapy for AIS.[60]	Prospective observational
Perioperative			
	Cardiac surgery	Intraoperative cerebral desaturations may be associated with more major organ morbidity and mortality.[34]	Randomized controlled trial
		The role of intraoperative cerebral desaturations in postoperative cognitive decline is unclear.[34,35,37]	Randomized controlled trial
		Intraoperative cerebral desaturations may be associated with protracted ICU[34] and hospital[35] LOS, although other authors disagree.[36]	Randomized controlled trial
	Carotid surgery	There are similar accuracy and reproducibility in the detection of cerebral ischemia compared with TCD and stump pressure.[39]	Prospective observational
	Head-up (beach chair) position surgery	Hypotension-associated decreases in rSco$_2$ are not associated with a higher incidence of postoperative cognitive dysfunction or serum biomarkers of brain injury.[61]	Prospective observational

Abbreviations: AIS, acute ischemic stroke; AVM, arteriovenous malformation; GOS, Glasgow Outcome Scale score; LOS, length of stay; MCA, middle cerebral artery; ORx, oxygen reactivity index; TCD, transcranial Doppler.

Box 1
The main variables known to influence brain tissue oxygen tension values

Systemic variables

- Pao_2
- $Paco_2$
- Fio_2
- Mean arterial blood pressure
- Cardiopulmonary function
- Hemoglobin level

Brain-specific variables

- CPP and ICP
- CBF
- Cerebral vasospasm
- Cerebral autoregulatory status
- Brain tissue gradients for oxygen diffusion
- Composition of the microvasculature around the probe and the relative dominance of arterial or venous vessels

---

Management of low brain tissue oxygen tension

There is currently no consensus on how low $Ptio_2$ should be treated. A stepwise approach has been recommended in a manner akin to the treatment of raised ICP, incorporating knowledge of the factors that influence $Ptio_2$ values (**Fig. 2**). Exactly which intervention, or combination of interventions, is most effective in improving $Ptio_2$ remains unclear. It seems that it is the responsiveness of the hypoxic brain to a given intervention that is the prognostic factor, with reversal of hypoxia associated with reduced mortality.[19]

Evidence for brain tissue oxygen tension–guided therapy on outcomes

There is a substantial body of evidence corroborating the relationship between low $Ptio_2$ values and adverse outcomes after TBI and SAH, but little for other conditions. Furthermore, although there is evidence that interventions, such as CPP augmentation, normobaric hyperoxia, and red blood cell transfusions, can improve low $Ptio_2$ values after ABI, robust evidence that this translates into improved outcomes is lacking. In TBI, most outcome-based studies of $Ptio_2$-guided therapy have compared standard ICP/CPP-guided therapy with $Ptio_2$-guided therapy in association with ICP/CPP-guided therapy, with conflicting findings (**Table 3**). Such variations in reported outcomes may be the result of heterogeneity in study design, including different patient populations, different thresholds for intervention, and the interventions used to treat low $Ptio_2$ as well as variable study endpoints. Despite difficulties in controlling for these variations, a systematic review of 4 studies incorporating 491 patients found overall outcome benefits from $Ptio_2$-directed therapy compared with ICP/CPP-guided therapy alone (odds ratio of favorable outcome = 2.1; 95% CI, 1.4 –3.1).[10] All studies included in this systematic review, however, were nonrandomized, and only 2 (with small sample sizes) were truly prospective.

Preliminary results have recently been released from a prospective, phase II randomized controlled brain tissue oxygen in TBI (Brain Tissue Oxygen Monitoring in

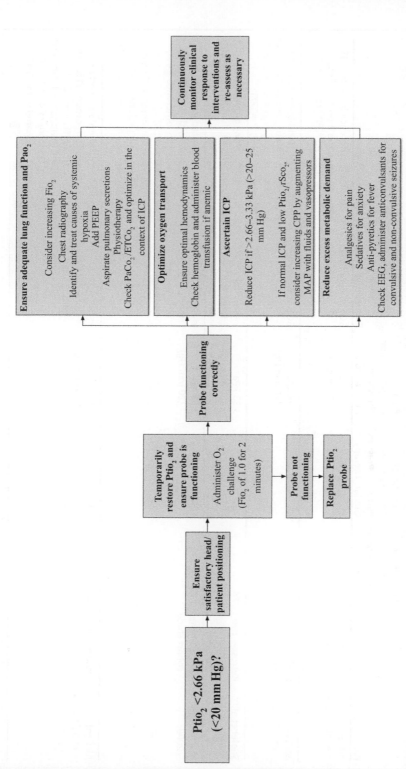

Fig. 2. Schematic for the management of low $Ptio_2$ values. EEG, electroencephalography; $ETco_2$, end-tidal carbon dioxide; MAP, mean arterial pressure; PEEP, positive end-expiratory pressure.

Table 3
Summary of studies comparing outcome after brain tissue oxygen tension–directed therapy versus intracranial pressure/cerebral perfusion pressure–directed therapy in patients with severe traumatic brain injury

Reference	Study Type	Brain Tissue Oxygen Tension–Guided Therapy, n	Intracranial Pressure/Cerebral Perfusion Pressure–Guided Therapy, n	Brain Tissue Oxygen Tension Threshold for Intervention	Endpoint(s)	Principal Findings
Adamides et al,[62] 2009	Prospective	20	10	2.66 kPa (20 mm Hg)	GOS at 6 months	No significant difference in mean GOS score between the 2 groups ($Ptio_2$ group = 3.55, ICP/CPP group = 4.40; $P = .19$)
Meixensberger et al,[63] 2003	Retrospective (historical controls)	53	40	1.33 kPa (10 mm Hg)	GOS at 6 months	A nonsignificant trend toward improved outcomes in the $Ptio_2$ group (GOS score of 4 or 5 in $Ptio_2$ group = 65%, in ICP/CPP group = 54%; $P = .27$)
Stiefel et al,[64] 2005	Retrospective (historical controls)	28	25	3.33 kPa (25 mm Hg)	In-hospital mortality	Mortality rate in $Ptio_2$ group (25%) significantly lower than ICP/CPP group (44%; $P<.05$)
Martini et al,[65] 2009	Retrospective cohort study	123	506	2.66 kPa (20 mm Hg)	In-hospital mortality, FIM at hospital discharge	Slightly worse adjusted mortality in $Ptio_2$ group compared with ICP/CPP group (adjusted mortality difference 4.4%, 95% CI, −3.9%–13%) Worse functional outcomes at discharge in the $Ptio_2$ group (adjusted FIM score difference −0.75; 95% CI, −1.41 to −0.09)
McCarthy et al,[66] 2009	Prospective	81	64	2.66 kPa (20 mm Hg)	In-hospital mortality, GOS every 3 mo postdischarge	No significant difference in mortality rates between $Ptio_2$ group (31%) and ICP/CPP group (36%; $P = .52$). A nonsignificant trend toward improved outcome (GOS score of 4 or 5) in $Ptio_2$ group (79%) compared with ICP/CPP group (61%; $P = .09$).

Study	Study design			Threshold	Outcome measures	Results
Narotam et al,[67] 2009	Retrospective (historical controls)	139	41	2.66 kPa (20 mm Hg)	GOS at 6 mo	Higher mean GOS score in $Ptio_2$ group (3.55 ± 1.75) compared with ICP/CPP group (2.71 ± 1.65; $P<.01$). OR for good outcome in $Ptio_2$ group = 2.09 (95% CI, 1.03–4.24). Reduced mortality rate in the $Ptio_2$ group (26% vs 41.5%; RR reduction 37%) despite higher ISS scores in $Ptio_2$ group
Spiotta et al,[68] 2010	Retrospective (historical controls)	70	53	2.66 kPa (20 mm Hg)	In-hospital mortality, GCS at 3 months	Mortality rates significantly lower in $Ptio_2$ group (26%) than ICP/CPP group (45%; $P<.05$). A favorable outcome (GOS score of 4 or 5) was also observed more commonly in the $Ptio_2$ group (64% vs 40%; $P = .01$).
Green et al,[69] 2013	Retrospective cohort study	37	37	2.66 kPa (20 mm Hg)	In-hospital mortality, GOS and FIM at discharge	No survival difference offered by $Ptio_2$-guided therapy (64.9% vs 54.1%, $P = .34$) or difference in discharge GCS or FIM. Of note, the $Ptio_2$-guided therapy group had significantly lower ISS at baseline.
Shutter,[20] 2014	Prospective phase II RCT	53	57	2.66 kPa (20 mm Hg)	Proportion of time $Ptio_2$ <20 mm Hg, safety, feasibility of protocol implementation, nonfutility (mortality, GOS-extended at 6 mo)	Median proportion of time with $Ptio_2$ <20 mm Hg significantly lower in the ICP/$Ptio_2$ group (0.14) compared with the ICP group (0.44; $P<.00001$). No significant difference between adverse events, and protocol violations were infrequent. Lower overall mortality and poor outcome on the 6-mo GOS-extended in the ICP/$Ptio_2$ group, but neither reached statistical significance ($P = .229$ and $P = .221$, respectively).

Abbreviations: FIM, functional independence measure; GCS, Glasgow Coma Scale; GOS, Glasgow Outcome Score; ISS, injury severity score; LOS, length of stay; MAP, mean arterial pressure; OR, odds ratio; RCT, randomized controlled trial; RR, relative risk.

Traumatic Brain Injury [BOOST]-2) trial, which evaluated the safety and efficacy of $Ptio_2$-directed therapy in 110 adult patients with nonpenetrating severe TBI.[20] Participants were randomized to receive either treatment based on ICP monitoring and management alone (target ICP <20 mm Hg) or treatment based on ICP (same target ICP) and $Ptio_2$ monitoring based on a prespecified protocol to maintain $Ptio_2$ greater than 20 mm Hg. The time spent with $Ptio_2$ less than 20 mm Hg was significantly lower in those in the ICP/$Ptio_2$ group, and there was no difference in adverse events between the 2 treatment arms. There was a trend toward lower overall mortality and less poor outcome in the ICP/$Ptio_2$ group, although these differences were not statistically significant ($P = .229$ and $P = .221$, respectively), which is unsurprising because this study was not powered for outcome. A larger phase III trial is required to clarify the potential outcome benefits of $Ptio_2$-guided therapy in TBI as well as in other brain injury types.

Near-infrared Spectroscopy

Near-infrared spectroscopy (NIRS)-derived cerebral oximetry is currently the only noninvasive, bedside monitor of cerebral oxygenation. Commercial devices measure regional cerebral oxygen saturation ($rSco_2$) with high temporal and spatial resolution and permit simultaneous measurement over multiple regions of interest. Despite interest in the clinical application of NIRS for more than 3 decades, widespread translation into routine clinical practice has not occurred.

Technical aspects

Full technical details of the principles of NIRS are beyond the scope of this review, and for further information readers are referred elsewhere.[21] In brief, NIRS systems are based on the transmission and absorption of near-infrared (NIR) light (wavelength range 700–950 nm) as it passes through tissue. Several biological molecules, termed *chromophores*, have distinct absorption spectra in the NIR, and their concentrations can be determined by their relative absorption of light in this wavelength range. From a clinical perspective, oxyhemoglobin and deoxyhemoglobin are the most commonly measured chromophores, although cytochrome-c oxidase (CCO), the terminal complex of the electron transfer chain, is increasingly investigated as a marker of cellular metabolism and may prove clinically more relevant.[22] In adults, NIR light cannot pass across the whole head so the light source and detecting devices are located a few centimeters apart on the same side of the head (reflectance spectroscopy), allowing examination of the superficial cortex.

NIRS interrogates arterial, venous, and capillary blood within the field of view, so $rSco_2$ values represent a weighted tissue oxygen saturation measured from these 3 compartments. $rSco_2$ values are also influenced by several physiologic variables, including arterial oxygen saturation, $Paco_2$, systemic blood pressure, hematocrit, cerebral blood flow (CBF), cerebral blood volume, cerebral metabolic rate for oxygen (CMo_2), and cerebral arterial:venous (a:v) ratio.[21]

Most commercial devices use spatially resolved spectroscopy to derive a scaled absolute hemoglobin concentration representing the relative proportions of oxyhemoglobin and deoxyhemoglobin within the field of view, from which $rSco_2$ is calculated and displayed as a percentage value. Frequency (or domain)-resolved spectroscopy and time-resolved spectroscopy allow measurement of absolute chromophore concentration with obvious advantages.[23] More recently, diffuse correlation spectroscopy techniques have been developed to monitor CBF and derive $CMRo_2$.[24,25]

There are several concerns over the clinical application of NIRS, in particular contamination of the signal by extracranial tissue. Some commercial cerebral oximeters use 2 detectors and a subtraction-based algorithm to deal with this problem, assuming that the detector closest to the emitter receives light that has passed mainly through the scalp and that the farthest light has passed mainly through brain tissue. Although there is weighting in favor of intracerebral tissue with an emitter-detector spacing greater than 4 cm, even spatially resolved spectroscopy is prone to some degree of extracerebral contamination,[26] and this is particularly problematic during low CBF states. The NIRS-derived CCO signal is highly specific for intracerebral changes, potentially making it a superior biomarker to hemoglobin-based NIRS variables.[27]

Indications

There are many Food and Drug Administration–approved NIRS devices available for clinical use. Because the specific algorithms incorporated into commercial devices vary, and are often unpublished, it is difficult to compare $rSco_2$ values between devices and, therefore, between studies using different devices. There are few numbers of high-quality clinical studies to guide the clinical use of NIRS, and prospective randomized studies are required not only to establish its potential role in patient monitoring but also to assess the relative efficacy of the multitude of devices on the market.[28]

Recent years have seen a significant increase in the use of perioperative cerebral oximetry (see **Table 2**), particularly during cardiac and carotid surgery, and surgery in the head-up (beach chair) position.[21] Although secondary ischemic injury after ABI is common, and low $rSco_2$ values have been associated with poor outcome in case series, data on the use of NIRS in the ICU management of ABI are limited and no outcome studies of NIRS-guided treatment have been published.

There are emerging applications for NIRS as a noninvasive monitor of cerebral autoregulation, using standard signal processing techniques[29] and also novel analytical techniques of multimodal monitoring of slow-wave oscillations.[30] NIRS has also been used to determine optimal CPP noninvasively in patients with TBI.[31] Because of the key role of CCO in mitochondrial metabolism, monitoring CCO concentration in addition to oxygenation variables may aid in the determination of ischemic thresholds after ABI by providing additional information about cellular metabolic status.[22]

One challenge in the application of NIRS after ABI is that the presence of intracranial hematoma, cerebral edema, or subarachnoid blood may invalidate some of the assumptions on which NIRS algorithms are based.[21] This has been used to advantage in the identification of intracranial hematomas and cerebral edema.[32,33]

Normal regional cerebral oxygen saturation values and thresholds for treatment

The normal range of $rSco_2$ is usually reported to lie between 60% and 75%, but there is substantial intraindividual and interindividual variability; NIRS-based cerebral oximetry is, therefore, best considered as a trend monitor. There are no validated $rSco_2$-derived ischemic thresholds,[21] but clinical studies and management protocols often use absolute $rSco_2$ values of less than or equal to 50% or a greater than or equal to 20% reduction from baseline as a trigger for initiating measures to improve cerebral oxygenation.

Evidence of near-infrared spectroscopy–guided therapy on outcomes

The only evidence for outcome effects of NIRS-guided treatment is from 3 randomized controlled trials in patients undergoing cardiac surgery[34–36] (**Table 4**). In small studies, intraoperative cerebral oxygen desaturation has been associated with early and late cognitive decline after cardiac surgery, but a recent systematic review concluded that only low-level evidence links intraoperative desaturation with postoperative neurologic complications.[37] There is also insufficient evidence to conclude that interventions to prevent or treat reductions in $rSco_2$ are effective in preventing stroke or postoperative cognitive dysfunction after cardiac surgery. NIRS-guided therapy may

Table 4
Summary of the randomized studies evaluating outcomes after near-infrared spectroscopy–guided management of cerebral desaturations

Reference	Surgery Type	Intervention arm, n	Control arm, n	Target Regional Cerebral Oxygen Saturation Values in Intervention Arm	Endpoint(s)	Principal Findings
Murkin et al,[34] 2007	CABG	100	100	≥75% of baseline threshold	30-d postoperative morbidity and mortality	Prolonged cerebral desaturations ($P = .014$) and ICU stay observed in the control group ($P = .029$) MOMM[a] higher in the control group ($P = .048$) and associated with lower baseline and mean $rSco_2$, more cerebral desaturations, longer ICU and hospital LOS
Slater et al,[35] 2009	CABG	125	115	≥80% of baseline threshold	Cognitive function (using a battery of tests) up to 3 mo and hospital LOS	A high $rSco_2$ desaturation score[b] (>3000% second) associated with early postoperative cognitive decline ($P = .024$) and hospital stay >6 d ($P = .007$) Benefits of the intervention on cognitive outcomes lost in multivariate analysis
Deschamps et al,[36] 2013	High-risk cardiac surgery[c]	23	25	≥80% of baseline threshold for 15 s	Desaturation load (% desaturation × time), ICU and hospital LOSs	Half of the patients had cerebral desaturations that could be reversed 88% of the time. Interventions resulted in smaller desaturation loads in the operating room and ICU. No difference in hospital or ICU LOS

Abbreviations: CABG, coronary artery bypass grafting; LOS, length of stay; OR, odds ratio.
[a] Major organ morbidity and mortality (MOMM) composed of the following variables as determined by the Society of Thoracic Surgeons: death, stroke, reoperation for bleeding, mediastinitis, surgical reintervention, renal failure requiring dialysis, and ventilation time greater than 48 hours.
[b] Desaturation score calculated by multiplying $rSco_2$ less than 50% by duration in seconds.
[c] High-risk cardiac surgery defined by the authors as redo surgery, adult congenital surgery, thoracic aortic surgery with and without circulatory arrest, and combined procedures surgery.

improve overall organ outcome after cardiac surgery, suggesting a role for NIRS as a monitor of overall organ perfusion.[34] A systematic review of the role of $rSco_2$ in pediatric patients undergoing surgery for congenital heart disease also concluded that there is no evidence that $rSco_2$ monitoring and management lead to a clinical improvement in short-term neurologic outcome in this patient population.[38]

Several methods are used to assess the adequacy of CBF and oxygen delivery during carotid surgery to inform the critical decision regarding shunt placement during the vessel cross-clamp period. Cerebral oximetry has similar accuracy for the detection of cerebral ischemia compared with other commonly used monitoring modalities and has advantages in terms of simplicity.[39] Various thresholds have been used to determine the need for shunt placement, ranging from a 10% to 20% reduction in ipsilateral $rSco_2$ from baseline.[40] Cerebral oximetry has higher temporal and spatial resolution compared with other modalities and thus may find a role in guiding the manipulation of systemic physiology to minimize the risk of cerebral hypoxia/ischemia during carotid surgery.[21] Recently, the use of a time-resolved optical imaging system, which allows simultaneous acquisition of data from 32 regions of interest over both hemispheres, has been described.[41] Distinct patterns of changes in hemoglobin and oxyhemoglobin were observed in ipsilateral brain cortex, suggesting that noninvasive optical imaging of brain tissue hemodynamics may find a role during carotid surgery.

There has been intense interest in the application of cerebral oximetry in patients undergoing surgery in the beach chair position because of the risk of hypotension-related cerebral ischemic events in anesthetized patients in the steep head-up position. In a recent study of 50 patients undergoing shoulder surgery in the beach chair position, the incidence of intraoperative cerebral desaturation events (defined as decreases in $rSco_2$ of $\geq 20\%$ from baseline) was 18%.[42] Of those experiencing desaturation, the mean maximal decrease in $rSco_2$ was 32% from preoperative baseline, and the mean number of separate desaturation events was 1.89 with an average duration of more than 3 minutes. Despite this apparently alarming high burden of cerebral desaturation, these authors and other investigators have not identified an association between desaturation events and postoperative neurocognitive dysfunction in this patient population. It has been suggested that changes in intracranial geometry and cerebral a:v ratio related to movement from supine to upright position might account, at least in part, for the changes in measured cerebral saturations.[23]

Finally, there is no evidence that monitoring and early detection of cerebral desaturations to guide targeted interventions improves perioperative outcomes during other surgical procedures under general anesthesia.[40]

FUTURE PERSPECTIVES

Technological developments are likely to be key drivers in advancing cerebral oxygenation monitoring and its adoption in the ICU and perioperative settings. A multi-parameter probe that combines ICP, $Ptio_2$, and temperature measurements is available commercially (Raumedic, Münchberg, Germany), allowing multimodality monitoring via a single invasive device. The addition of CBF quantification into such a probe is likely. Advances in $Ptio_2$ technology should allow for improved insertion techniques and more durable devices, and stereotactic placement of invasive probes may help target regions of interest with improved accuracy.

Cerebral arterial oxygen saturation has been estimated using fiberoptic pulse oximetry,[43] and a prototype invasive probe that combines NIRS and indocyanine green dye dilution has been investigated for the simultaneous monitoring of ICP,

CBF, and cerebral blood volume, avoiding NIRS signal contamination by extracerebral tissues.[44] Combined NIRS/electroencephalography provides a unique opportunity to acquire, noninvasively and simultaneously, regional cerebral electrophysiologic and hemodynamic data to elucidate on neurovascular coupling mechanisms.[45] A prototype device combining diffuse correlation spectroscopy and NIRS for the bedside measurement of CBF and cerebral oxygenation respectively has been described.[46] In the future, a single NIRS-based device may be able to provide noninvasive monitoring of cerebral oxygenation, hemodynamics, and cellular metabolic status over multiple regions of interest, although substantial technological advances are necessary before any of these techniques can be introduced into routine clinical practice.

SUMMARY

Cerebral oxygenation represents the balance between cerebral oxygen supply and demand, and a mismatch may lead to cerebral hypoxia/ischemia with deleterious outcomes. There are several tools available for the detection of cerebral hypoxia/ischemia, each with inherent advantages and disadvantages. Although there is a large body of evidence supporting an association between cerebral hypoxia/ischemia and poor outcomes, it remains to be determined whether restoring cerebral oxygenation improves outcomes. The adoption of a multimodality neuromonitoring approach (see Kirkman MA, Smith M: Multimodality Neuromonitoring, in this issue) that incorporates cerebral oxygenation monitoring in addition to more established ICP and CPP monitoring is required, along with large randomized prospective studies to address the current uncertainties about such approaches.

REFERENCES

1. Duckworth JL, Stevens RD. Imaging brain trauma. Curr Opin Crit Care 2010;16: 92–7.
2. Schell RM, Cole DJ. Cerebral monitoring: jugular venous oximetry. Anesth Analg 2000;90:559–66.
3. Macmillan CS, Andrews PJ. Cerebrovenous oxygen saturation monitoring: practical considerations and clinical relevance. Intensive Care Med 2000;26: 1028–36.
4. Rose JC, Neill TA, Hemphill JC. Continuous monitoring of the microcirculation in neurocritical care: an update on brain tissue oxygenation. Curr Opin Crit Care 2006;12:97–102.
5. Bouzat P, Sala N, Payen J-F, et al. Beyond intracranial pressure: optimization of cerebral blood flow, oxygen, and substrate delivery after traumatic brain injury. Ann Intensive Care 2013;3:23.
6. Oddo M, Levine JM, Mackenzie L, et al. Brain hypoxia is associated with short-term outcome after severe traumatic brain injury independently of intracranial hypertension and low cerebral perfusion pressure. Neurosurgery 2011;69: 1037–45.
7. Kirkman MA, Smith M. Intracranial pressure monitoring, cerebral perfusion pressure estimation, and ICP/CPP-guided therapy: a standard of care or optional extra after brain injury? Br J Anaesth 2014;112:35–46.
8. Muench E, Horn P, Bauhuf C, et al. Effects of hypervolemia and hypertension on regional cerebral blood flow, intracranial pressure, and brain tissue oxygenation after subarachnoid hemorrhage. Crit Care Med 2007;35:1844–51.

9. van den Brink WA, van Santbrink H, Steyerberg EW, et al. Brain oxygen tension in severe head injury. Neurosurgery 2000;46:868–76.

10. Nangunoori R, Maloney-Wilensky E, Stiefel M, et al. Brain tissue oxygen-based therapy and outcome after severe traumatic brain injury: a systematic literature review. Neurocrit Care 2012;17:131–8.

11. Oddo M, Bösel J, Participants in the International Multidisciplinary Consensus Conference on Multimodality Monitoring. Monitoring of brain and systemic oxygenation in neurocritical care patients. Neurocrit Care 2014;21:103–20.

12. Brain Trauma Foundation. Guidelines for the management of severe traumatic brain injury. J Neurotrauma 2007;24(Suppl 1):S1–106.

13. Jaeger M, Schuhmann MU, Soehle M, et al. Continuous monitoring of cerebro-vascular autoregulation after subarachnoid hemorrhage by brain tissue oxygen pressure reactivity and its relation to delayed cerebral infarction. Stroke 2007;38:981–6.

14. Kirkman MA, Smith M. Supratentorial intracerebral hemorrhage: a review of the underlying pathophysiology and its relevance for multimodality neuromonitoring in neurointensive care. J Neurosurg Anesthesiol 2013;25:228–39.

15. Diringer MN, Bleck TP, Claude Hemphill J, et al. Critical care management of patients following aneurysmal subarachnoid hemorrhage: recommendations from the Neurocritical Care Society's Multidisciplinary Consensus Conference. Neurocrit Care 2011;15:211–40.

16. Rosenthal G, Hemphill JC, Sorani M, et al. Brain tissue oxygen tension is more indicative of oxygen diffusion than oxygen delivery and metabolism in patients with traumatic brain injury. Crit Care Med 2008;36:1917–24.

17. Johnston AJ, Steiner LA, Coles JP, et al. Effect of cerebral perfusion pressure augmentation on regional oxygenation and metabolism after head injury. Crit Care Med 2005;33:189–95.

18. Damiani E, Adrario E, Girardis M, et al. Arterial hypoxia and mortality in critically ill patients: a systematic review and meta-analysis. Crit Care 2014;18:711.

19. Bohman L-E, Pisapia JM, Sanborn MR, et al. Response of brain oxygen to therapy correlates with long-term outcome after subarachnoid hemorrhage. Neurocrit Care 2013;19:320–8.

20. Shutter L. BOOST 2 Trial Study Results. Annual Neurocritical Care Society Meeting. Available at: http://www.neurocriticalcare.org/news/2014-annual-meeting-highlights. Accessed October 13, 2015.

21. Ghosh A, Elwell C, Smith M. Review article: cerebral near-infrared spectroscopy in adults: a work in progress. Anesth Analg 2012;115:1373–83.

22. Smith M, Elwell C. Near-infrared spectroscopy: shedding light on the injured brain. Anesth Analg 2009;108:1055–7.

23. Smith M. Shedding light on the adult brain: a review of the clinical applications of near-infrared spectroscopy. Philos Trans A Math Phys Eng Sci 2011;369:4452–69.

24. Diop M, Verdecchia K, Lee TY, et al. Calibration of diffuse correlation spectros-copy with a time-resolved near-infrared technique to yield absolute cerebral blood flow measurements. Biomed Opt Express 2011;2:2068–81.

25. Verdecchia K, Diop M, Lee T-Y, et al. Quantifying the cerebral metabolic rate of oxygen by combining diffuse correlation spectroscopy and time-resolved near-infrared spectroscopy. J Biomed Opt 2013;18:27007.

26. Davie SN, Grocott HP. Impact of extracranial contamination on regional cerebral oxygen saturation: a comparison of three cerebral oximetry technologies. Anes-thesiology 2012;116:834–40.

27. Kolyva C, Ghosh A, Tachtsidis I, et al. Cytochrome c oxidase response to changes in cerebral oxygen delivery in the adult brain shows higher brain-specificity than haemoglobin. Neuroimage 2014;85(Pt 1):234–44.
28. Douds M, Straub E, Kent A, et al. A systematic review of cerebral oxygenation-monitoring devices in cardiac surgery. Perfusion 2014;29(6):545–52.
29. Zweifel C, Castellani G, Czosnyka M, et al. Noninvasive monitoring of cerebrovascular reactivity with near infrared spectroscopy in head-injured patients. J Neurotrauma 2010;27:1951–8.
30. Highton D, Ghosh A, Tachtsidis I, et al. Monitoring cerebral autoregulation after brain injury: multimodal assessment of cerebral slow-wave oscillations using near-infrared spectroscopy. Anesth Analg 2015;121:198–205.
31. Dias C, Silva MJ, Pereira E, et al. Optimal cerebral perfusion pressure management at bedside: a single-center pilot study. Neurocrit Care 2015;23:92–102.
32. Gopinath SP, Robertson CS, Contant CF, et al. Early detection of delayed traumatic intracranial hematomas using near-infrared spectroscopy. J Neurosurg 1995;83:438–44.
33. Damian MS, Schlosser R. Bilateral near infrared spectroscopy in space-occupying middle cerebral artery stroke. Neurocrit Care 2007;6:165–73.
34. Murkin JM, Adams SJ, Novick RJ, et al. Monitoring brain oxygen saturation during coronary bypass surgery: a randomized, prospective study. Anesth Analg 2007; 104:51–8.
35. Slater JP, Guarino T, Stack J, et al. Cerebral oxygen desaturation predicts cognitive decline and longer hospital stay after cardiac surgery. Ann Thorac Surg 2009; 87:36–44 [discussion: 44–5].
36. Deschamps A, Lambert J, Couture P, et al. Reversal of decreases in cerebral saturation in high-risk cardiac surgery. J Cardiothorac Vasc Anesth 2013;27: 1260–6.
37. Zheng F, Sheinberg R, Yee M-S, et al. Cerebral near-infrared spectroscopy monitoring and neurologic outcomes in adult cardiac surgery patients: a systematic review. Anesth Analg 2013;116:663–76.
38. Hirsch JC, Charpie JR, Ohye RG, et al. Near-infrared spectroscopy: what we know and what we need to know–a systematic review of the congenital heart disease literature. J Thorac Cardiovasc Surg 2009;137:154–9, 159.e1–12.
39. Moritz S, Kasprzak P, Arlt M, et al. Accuracy of cerebral monitoring in detecting cerebral ischemia during carotid endarterectomy: a comparison of transcranial Doppler sonography, near-infrared spectroscopy, stump pressure, and somatosensory evoked potentials. Anesthesiology 2007;107:563–9.
40. Nielsen HB. Systematic review of near-infrared spectroscopy determined cerebral oxygenation during non-cardiac surgery. Front Physiol 2014;5:93.
41. Kacprzak M, Liebert A, Staszkiewicz W, et al. Application of a time-resolved optical brain imager for monitoring cerebral oxygenation during carotid surgery. J Biomed Opt 2012;17:016002.
42. Salazar D, Sears BW, Andre J, et al. Cerebral desaturation during shoulder arthroscopy: a prospective observational study. Clin Orthop Relat Res 2013; 471:4027–34.
43. Phillips JP, Langford RM, Chang SH, et al. Cerebral arterial oxygen saturation measurements using a fiber-optic pulse oximeter. Neurocrit Care 2010;13: 278–85.
44. Keller E, Froehlich J, Muroi C, et al. Neuromonitoring in intensive care: a new brain tissue probe for combined monitoring of intracranial pressure (ICP) cerebral blood flow (CBF) and oxygenation. Acta Neurochir Suppl 2011;110:217–20.

45. Cooper RJ, Hebden JC, O'Reilly H, et al. Transient haemodynamic events in neurologically compromised infants: a simultaneous EEG and diffuse optical imaging study. Neuroimage 2011;55:1610–6.
46. Kim MN, Durduran T, Frangos S, et al. Noninvasive measurement of cerebral blood flow and blood oxygenation using near-infrared and diffuse correlation spectroscopies in critically brain-injured adults. Neurocrit Care 2010;12:173–80.
47. Eriksson EA, Barletta JF, Figueroa BE, et al. The first 72 hours of brain tissue oxygenation predicts patient survival with traumatic brain injury. J Trauma Acute Care Surg 2012;72:1345–9.
48. Meixensberger J, Renner C, Simanowski R, et al. Influence of cerebral oxygenation following severe head injury on neuropsychological testing. Neurol Res 2004; 26:414–7.
49. Radolovich DK, Czosnyka M, Timofeev I, et al. Transient changes in brain tissue oxygen in response to modifications of cerebral perfusion pressure: an observational study. Anesth Analg 2010;110:165–73.
50. Bohman L-E, Heuer GG, Macyszyn L, et al. Medical management of compromised brain oxygen in patients with severe traumatic brain injury. Neurocrit Care 2011;14:361–9.
51. Ramakrishna R, Stiefel M, Udoetuk J, et al. Brain oxygen tension and outcome in patients with aneurysmal subarachnoid hemorrhage. J Neurosurg 2008;109: 1075–82.
52. Jaeger M, Soehle M, Schuhmann MU, et al. Clinical significance of impaired cerebrovascular autoregulation after severe aneurysmal subarachnoid hemorrhage. Stroke 2012;43:2097–101.
53. Ko S-B, Choi HA, Parikh G, et al. Multimodality monitoring for cerebral perfusion pressure optimization in comatose patients with intracerebral hemorrhage. Stroke 2011;42:3087–92.
54. Dohmen C, Bosche B, Graf R, et al. Identification and clinical impact of impaired cerebrovascular autoregulation in patients with malignant middle cerebral artery infarction. Stroke 2007;38:56–61.
55. Carvi y Nievas M, Toktamis S, Höllerhage HG, et al. Hyperacute measurement of brain-tissue oxygen, carbon dioxide, pH, and intracranial pressure before, during, and after cerebral angiography in patients with aneurysmatic subarachnoid hemorrhage in poor condition. Surg Neurol 2005;64:362–7.
56. Jödicke A, Hübner F, Böker D-K. Monitoring of brain tissue oxygenation during aneurysm surgery: prediction of procedure-related ischemic events. J Neurosurg 2003;98:515–23.
57. Ibanez J, Vilalta A, Mena MP, et al. Intraoperative detection of ischemic brain hypoxia using oxygen tissue pessure microprobes (in Spanish). Neurocirugia (Astur) 2003;14:483–9.
58. Dunham CM, Ransom KJ, Flowers LL, et al. Cerebral hypoxia in severely brain-injured patients is associated with admission Glasgow Coma Scale score, computed tomographic severity, cerebral perfusion pressure, and survival. J Trauma 2004;56:482–9 [discussion: 489–91].
59. Yokose N, Sakatani K, Murata Y, et al. Bedside monitoring of cerebral blood oxygenation and hemodynamics after aneurysmal subarachnoid hemorrhage by quantitative time-resolved near-infrared spectroscopy. World Neurosurg 2010;73:508–13.
60. Hametner C, Stanarcevic P, Stampfl S, et al. Noninvasive cerebral oximetry during endovascular therapy for acute ischemic stroke: an observational study. J Cereb Blood Flow Metab 2015;35(11):1722–8.

61. Laflam A, Joshi B, Brady K, et al. Shoulder surgery in the beach chair position is associated with diminished cerebral autoregulation but no differences in postoperative cognition or brain injury biomarker levels compared with supine positioning: the anesthesia patient safety foundation beach chair study. Anesth Analg 2015;120:176–85.

62. Adamides AA, Cooper DJ, Rosenfeldt FL, et al. Focal cerebral oxygenation and neurological outcome with or without brain tissue oxygen-guided therapy in patients with traumatic brain injury. Acta Neurochir (Wien) 2009;151:1399–409.

63. Meixensberger J, Jaeger M, Vath A, et al. Brain tissue oxygen guided treatment supplementing ICP/CPP therapy after traumatic brain injury. J Neurol Neurosurg Psychiatr 2003;74:760–4.

64. Stiefel MF, Spiotta A, Gracias VH, et al. Reduced mortality rate in patients with severe traumatic brain injury treated with brain tissue oxygen monitoring. J Neurosurg 2005;103:805–11.

65. Martini RP, Deem S, Yanez ND, et al. Management guided by brain tissue oxygen monitoring and outcome following severe traumatic brain injury. J Neurosurg 2009;111:644–9.

66. McCarthy MC, Moncrief H, Sands JM, et al. Neurologic outcomes with cerebral oxygen monitoring in traumatic brain injury. Surgery 2009;146:585–90.

67. Narotam PK, Morrison JF, Nathoo N. Brain tissue oxygen monitoring in traumatic brain injury and major trauma: outcome analysis of a brain tissue oxygen-directed therapy. J Neurosurg 2009;111:672–82.

68. Spiotta AM, Stiefel MF, Gracias VH, et al. Brain tissue oxygen-directed management and outcome in patients with severe traumatic brain injury. J Neurosurg 2010;113:571–80.

69. Green JA, Pellegrini DC, Vanderkolk WE, et al. Goal directed brain tissue oxygen monitoring versus conventional management in traumatic brain injury: an analysis of in hospital recovery. Neurocrit Care 2013;18:20–5.

Neurocritical Care:
Michael L. "Luke" James

Controversies in the Management of Traumatic Brain Injury

Sayuri Jinadasa, MD, M. Dustin Boone, MD*

KEYWORDS

- Traumatic brain injury • Cerebral autoregulation
- Traumatic brain injury management • Traumatic brain injury outcomes

KEY POINTS

- In the United States, an estimated 1.7 million people sustain a traumatic brain injury annually, with an annual mortality of 52,000 people.
- The management of traumatic brain injury has been standardized with the publication of the Brain Trauma Foundation Guidelines; however, much of the evidence for these guidelines is poor or conflicting.
- One of the main aims of treatment of traumatic brain injury is to prevent or treat increased intracranial pressure.

EPIDEMIOLOGY

Often referred to as the silent epidemic, the after-effects of traumatic brain injury (TBI), including changes in cognition, sensation, language, and emotions, are not always readily apparent. In addition, the general public's awareness about TBI seems limited,[1] and long-term disability after TBI may be overlooked.[2]

TBI is a worldwide source of significant social and health care burden that results from assaults, motor vehicle accidents, falls, sports collisions, and combat-related injuries. In the United States alone, an estimated 1.7 million people sustain a TBI annually, with an annual mortality of 52,000 people.[1] In 2003, 124,626 people experienced long-term disability following TBI,[3] some requiring lifelong assistance to perform activities of daily living.[2] On average, the direct medical cost for severe TBI was $65,600 per patient in 2002.[4]

This work is published in collaboration with the Society for Neuroscience in Anesthesiology and Critical Care.
Conflicts of Interest: Neither author has any direct or indirect commercial or financial incentives associated with publishing this article.
Department of Anesthesia, Beth Israel Deaconess Medical Center, Harvard Medical School, 330 Brookline Avenue, Boston, MA 02215, USA
* Corresponding author. Department of Anesthesia, Beth Israel Deaconess Medical Center, 1 Deaconess Road, Boston, MA 02215.
E-mail address: mboone@bidmc.harvard.edu

Anesthesiology Clin 34 (2016) 557–575
http://dx.doi.org/10.1016/j.anclin.2016.04.008
1932-2275/16/$ – see front matter © 2016 Elsevier Inc. All rights reserved.

Men have greater incidence of TBI compared to women,[5] presumably related to more risky behavior in men. Although TBI is particularly prevalent in the younger population, the average age of TBI in the United States is increasing with the overall age of the population. According to the Centers for Disease Control and Prevention, approximately 81% of TBIs in senior citizens are caused by falls.[1]

PATHOPHYSIOLOGY
Autoregulation

Cerebral autoregulation (CA) is the brain's mechanism of maintaining adequate and stable blood flow despite changes in perfusion pressure. As mean arterial pressure (MAP) changes, cerebral vascular resistance changes to maintain constant flow (**Fig. 1**) via arteriolar vasodilation and constriction, consistent with Poiseuille's law for laminar flow through a vessel:

$$\text{Flow} = \frac{\pi\ (pressure\ difference)\ (r)^4}{8\ (viscosity)\ (length)}$$

CA applies not only to the brain as a whole; regional areas of high metabolic demand and neuronal activity will receive greater blood flow.[6]

Cerebral blood flow is maintained over a range of a cerebral perfusion pressure of 60 to 160 mm Hg (cerebral perfusion pressure is the total of intracranial pressure [ICP] subtracted from MAP). Above and below this range, CA is lost, and blood flow depends on MAP in a linear fashion (see **Fig. 1**). In this setting, if cerebral blood flow decreases because of hypotension, compensation includes increased oxygen extraction. Once this compensatory mechanism is exhausted, cerebral ischemia and infarction can occur. Alternatively, if cerebral blood flow increases because of hypertension, vascular constriction is overcome by excessive intravascular pressure. This increased pressure experienced by the cerebral endothelium may lead to edema formation and eventual herniation.

CA is often impaired in severe TBI. Thus, techniques to evaluate the integrity of the autoregulatory mechanism have been extensively studied. One such method involves using a linear correlation coefficient between arterial blood pressure and ICP. This

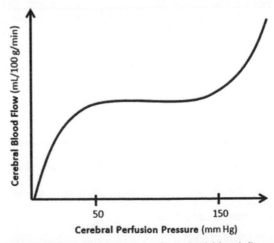

Fig. 1. Cerebral autoregulation. Maintenance of cerebral blood flow by autoregulation typically occurs within a MAP range of 60 to 160 mm Hg.

method is known as the Cerebrovascular Pressure Reactivity Index.[7] Negative values suggest that CA is intact, whereas positive values suggest impairment (essentially ICP trends with MAP). As clinicians recognize the heterogeneity of TBI, there is increasing interest in individualizing management. With knowledge of the integrity of CA and information from advanced multimodality monitoring, it may be possible to individualize therapy. For example, with intact CA, a decision could be made to increase the ICP threshold that would otherwise prompt intervention. Because no large-scale trial has evaluated TBI outcomes using this approach, multimodality monitoring remains experimental.

Phases of Brain Injury

There are 2 pathophysiologic phases of neurologic damage in TBI: primary and secondary injury. Primary injury refers to initial damage occurring at the moment of impact, which involves mechanical forces that shear and compress neurons, glia, and vascular tissue. This initial damage leads to physical disruption of cell membranes and their infrastructure with subsequent increase in membrane permeability and disturbance of ionic homeostasis. This initial damage in turn leads to cellular swelling, relative hypoperfusion, and a cascade of neurotoxic events.[8–13]

Secondary injury that occurs in TBI is an escalating cyclical positive feedback process of cerebral edema, increasing ICP, decreasing cerebral blood flow and oxygen supply, and cell death.[14] This secondary injury, which occurs in the ensuing hours and days after the primary insult, leads to physiologic consequences such as brain tissue ischemia and reperfusion injury.[15] Cerebral edema results in increased ICP, which can cause mechanical damage by cerebral herniation or by impingement of cerebral blood flow that then results in ischemia. This ischemia may be exacerbated by systemic hypotension and hypoxia experienced before resuscitation.[16–18]

CLASSIFICATION

Classification schemas allow succinct communication about the severity of a patient's status in a standardized fashion. They allow the tracking of patients' clinical progress and provide a means of prognostication, which is important to guide clinical decision making. Although many ways to classify TBI exist, the following are the most commonly used.

Head Injury Classification Based on Clinical Examination

Glasgow Coma Scale

The Glasgow Coma Scale (GCS) is an assessment tool used to objectively characterize neurologic status, ranging from scores of 3 to 15 based on the best eye, verbal, and motor responses (**Table 1**). Developed at the University of Glasgow in 1974, the scale standardizes observations of level of consciousness in patients with brain damage caused by traumatic injury, vascular insults, infection, metabolic disorders, or drug overdose.[19] It has been subjected to extensive interrater and intrarater reliability testing and is now widely accepted and used.[20–22] This scale has especially high utility because it can be graded rapidly with no need for diagnostic equipment. A patient's initial GCS score after resuscitation conveys prognostic information about mortality and functional outcomes at 6 months.[23] There is a linear relationship with a poor outcome, including death, vegetative state, or severe neurologic disability, in the GCS range of 3 to 9,[24] although there are notable exceptions to this. For example, patients with head injury who have epidural hematomas and display the "talk and die" syndrome initially have no deficits (the lucid interval), but eventually lapse into a

Table 1 GCS		
E	**V**	**M**
4 = Spontaneous	5 = Oriented	6 = Obeys commands
3 = To speech	4 = Disoriented	5 = Localized to pain
2 = To pain	3 = Inappropriate	4 = Withdraws from pain
1 = None	2 = Incomprehensible	3 = Decorticate posture
	1 = None	2 = Decerebrate posture
		1 = None
CGS = E + V + M Minimum score = 3 Maximum score = 15		

Abbreviations: E, eye opening; M, motor response; V, verbal response.

coma and may suffer herniation and death if not detected within an appropriate time period.

Patients with a GCS score between 13 and 15 are considered to have a mild TBI, usually a concussion. Despite potential short-term memory loss or difficulty concentrating, full neurologic recovery is expected. A GCS score between 9 and 12 is considered a moderate TBI, and patients may initially be lethargic or stuporous. Severe TBI is classified as a GCS between 3 and 8, meaning that patients are comatose.[24] Any patient with a GCS of 8 or lower after TBI has a high likelihood of requiring mechanical ventilation for airway protection. Deterioration of a patient's GCS score may indicate a developing cerebral bleed that may require surgical evacuation, increased edema, or medication effects. It is important to consider that patients with TBI may have abnormal GCS scores that are not only altered because of the head injury itself but also because of hypotension, hypoxemia, or ingestion of neurologic depressants.

The adult GCS can be used in patients 5 years of age or older. The pediatric scale is modified to account for the immaturity of the pediatric nervous system.[15]

There are multiple shortcomings associated with the scale, including interrater disagreement,[20,22] failure to incorporate brainstem reflexes,[25] and inability to accurately calculate the GCS in intubated, sedated, paralyzed, and immobilized patients, and those with significant periorbital swelling. Despite these shortcomings, the GCS score remains the most utilized level-of-consciousness scale worldwide.[25]

Pupil reactivity
The pupillary examination is a key component of the neurologic assessment. A pupil examination includes assessment of the pupil's size and equality, shape, and reactivity to light. It is imperative to be aware of any ophthalmologic procedures a patient has undergone because these may affect the examination. For example, pupillary constriction is reduced after cataract surgery.

Reactivity to light is assessed in the ipsilateral and contralateral pupils, and they are recorded as equally round and reactive to light, unilaterally or bilaterally nonreactive, fixed, and/or dilated. The rate at which pupils react is described as brisk or sluggish.

In patients with TBI, the pupillary examination may at first be benign but can change with evolving intracranial pathology hours after the primary injury. Serial pupillary examinations are crucial to detect subtle changes in the patient's neurologic status. The pupillary examination is particularly useful because it is one of the few neurologic examinations that are noninvasive and that can be performed on unconscious, sedated, and paralyzed patients.[26] **Fig. 2** shows various cerebral lesions and the resulting pupillary examination.

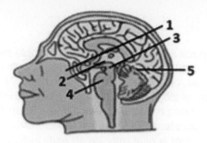

1. Diencephalic lesion

2. Third nerve impingement (uncal herniation)

3. Midbrain lesion

4. Pontine lesion

5. Tectal lesion

Fig. 2. Pupillary examination. Cerebral lesions and the resulting pupillary examinations. (*Adapted from* Seder D, Mayer SA, Frontera JA. Management of elevated intracranial pressure. In: Frontera JA, editor. Decision making in neurocritical care. New York: Thieme Medical Publishers; 2009. p. 201; with permission.)

Head injury can affect pupil size and reactivity via multiple mechanisms, including direct injury to the eye and optic nerve, cranial nerve (CN) III injury or compression along any point of its course, or midbrain and pontine hypoperfusion.[27–32]

The pupillary reflexes travel along the oculomotor nerve, which is sensitive to mass effect such as brain stem compression and ischemia. When CN III is compressed because of an increase in supratentorial pressure, ipsilateral mydriasis and loss of light reactivity ensues.[33] A pupil that is oval may be an early indication of compression of CN III caused by increased ICP. Nonreactive pupils are often associated with severe increases in ICP and/or severe brain damage/herniation. Thus, an abnormal or changed pupillary examination in a patient with TBI can provide valuable information regarding deterioration in ICP from edema and any expanding hematomas. Pupillary changes also occur with disruptions in brainstem oxygenation and perfusion, and may not always indicate herniation.[34]

Chen and colleagues[35] created a pupil index based on the pupillary examination performed using a portable hand-held pupillometer. They found an inverse trend in the relationship between decreasing pupil reactivity and increasing ICP and that pupil reactivity is an early indicator of increasing ICP in patients with TBI, aneurysmal subarachnoid hemorrhage, or intracranial hemorrhage. Patients with an abnormal pupillary light reactivity had an average peak ICP of 30.5 mm Hg versus 19.6 mm Hg in the normal pupil reactivity population. Patients with nonreactive pupils had the highest mean ICP of 33.8 mm Hg. In the group of patients with abnormal pupillary reactivity, the first evidence of pupil abnormality occurred on average 15.9 hours before the time of peak ICP. This study indicates that the pupillary examination can provide useful predictive information even in patients with an indwelling ICP monitor.

The pupillary examination has prognostic significance. Studies have found that patients with severe TBI with postresuscitation bilaterally fixed pupils (<4 mm) were associated with a 70% to 90% chance of death or vegetative state, whereas patients whose postresuscitation pupil examination showed bilaterally reactive pupils had a 30% chance of poor outcome.[24,27–32] Multiple analyses have shown that a combination of pupil reactivity and the motor component of the GCS outmatches the predictive accuracy of the GCS score alone.[23,36,37]

Some of the drawbacks of the manual pupillary examination include that the interreader agreement rate is low at 39%,[38] physician or nurse measurements may be imprecise,[39] early changes in pupillary function may not be detectable to the naked eye,[40] and the examination may not be measurable in patients with extensive facial trauma. Given the critical prognostic information offered by the pupil examination combined with the poor interreader reliability and subjectivity of the bedside manual pupillary examination, the pupillometer has been introduced as an infrared device that provides a quantitative and reliable examination.[40] Measurements made by the pupillometer include pupil size, latency, constriction velocity, and dilation velocity. Studies have found that the pupillometer is more accurate and reliable than physician and nurse estimates of pupillary size and that subtler changes in a patient's pupillary examination can be detected earlier compared with the manual examination.[39,40] For these reasons, such a device, although not currently designated as standard of care, could provide an advancement in TBI management.

Head Injury Classification Based on Imaging

Marshall Classification
Certain computed tomography (CT) scan findings are ominous for developing intracranial hypertension and eventual catastrophic outcomes, particularly in patients who did

not appear to be severely injured based on their clinical examination.[41,42] Thus, the Marshall Criteria for categorizing diffuse injury in patients with moderate to severe TBI from a nonpenetrating injury were developed in 1991 and have become the de facto standard for CT classification.[15]

The grading system applies to CT scans performed in patients with nonfocal injury and accounts for the status of the mesencephalic cisterns, the degree of midline shift in millimeters, and the presence or absence of 1 or more surgical masses (**Fig. 3**, **Table 2**). Each category of diffuse injury has a progressively higher risk for intracranial hypertension and death. Patients with both severe TBI and abnormality detected on head CT on admission have a greater than 50% chance of intracranial hypertension.[24] A predictive model showed that a patient's Marshall CT classification combined with age and postresuscitation GCS motor score has an excellent direct relationship with mortality.[41,42] This risk association applies to CT scans performed during or after resuscitation, usually within 4 hours of injury.

Shortcomings of the Marshall system include significant interobserver variability[28,43] and the lack of a normal imaging category, such as in individuals with pathologically mild head injury but low GCS (eg, resulting from intoxication).

As the availability and speed of CT scanning improves, there is increasing risk of missing operable lesions in the initial CT scan; thus, serial imaging should be considered, and optimal timing for prognostic CT imaging is yet to be determined.

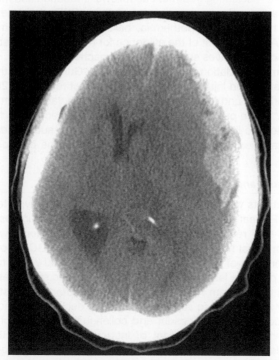

Fig. 3. Head CT scan showing compression of the lateral ventricles, midline shift greater than 5 mm, and a mixed-density lesion greater than 25 cm³. This CT scan has a Marshall Classification of nonevacuated mass.

Table 2
Marshall Classification of CT head findings

Category	Definition
Diffuse injury I	No visible intracranial disorder
Diffuse injury II	Cisterns present Midline shift <5 mm And/or no high-density or mixed-density lesion >25 cm³
Diffuse injury III	Cisterns are compressed or absent and Midline shift is 0–5 mm with no high-density or mixed-density lesion >25 cm³
Diffuse injury IV	Midline shift >5 mm with no high-density or mixed-density lesion >25 cm³
Evacuated mass	Any lesion surgically evacuated
Nonevacuated mass	High-density or mixed-density lesion >25 cm³ not surgically evacuated

MANAGEMENT

The overarching goal of treatment in patients with TBI is to minimize the progression of secondary brain injury, namely ischemia, by reducing cerebral edema and ICP while maintaining cerebral perfusion, oxygen delivery, and energy to the brain.[14] Simply put, the goal is to squeeze oxygenated blood through a swollen brain.[24]

Management of increased ICP has evolved toward standardized strategies that use a tiered approach of escalating treatment as follows: intubation and mechanical ventilation, sedation and analgesia, cerebrospinal fluid (CSF) drainage, hyperosmolar agents, induced hypocapnia, hypothermia, barbiturates, and decompressive craniotomy (**Fig. 4**).[44] Although most of the evidence for these treatment strategies is level II at best, expert opinion drives some of the approaches to care.

The Brain Trauma Foundation (BTF) developed the "Guidelines for the Management of Severe Traumatic Brain Injury," first in 1995 and most recently in 2007, to improve mortality and morbidity and to address the considerable variability in the care of patients with TBI.[45–47] The publication reviews evidence-based methodologies and the quality of the scientific literature validating them. Despite such guidelines, many of the treatment strategies remain highly controversial, and the best practice is yet to be determined.

Intracranial Pressure and Multimodal Monitoring

Up to 50% of patients with severe TBI may have intracranial hypertension, which is defined as an ICP greater than 20 mm Hg for longer than 5 minutes. Based on this observation, the BTF recommended monitoring ICP for patients with a GCS less than 8 and an abnormal head CT or for those patients with GCS 8 or less, a normal head CT, and 2 or more of the following: age greater than 40 years, motor posturing, or systolic blood pressure less than 90 mm Hg. Therapeutic strategies are directed at keeping ICP less than 20 mm Hg. Although widely adopted as the standard of care, there are notably limited data to support therapeutic decisions based on ICP.

A recent trial conducted by Chesnut and colleagues,[48] the Benchmark Evidence from South American Trials: Treatment of Intracranial Pressure (BEST:TRIP), cast doubt on the relative value of ICP monitoring compared with frequent neurologic examinations and CT imaging. They showed no difference in outcome between groups and concluded that a strategy to maintain ICP at less than 20 mm Hg in patients with an ICP monitor in place was not superior to a treatment strategy that relied

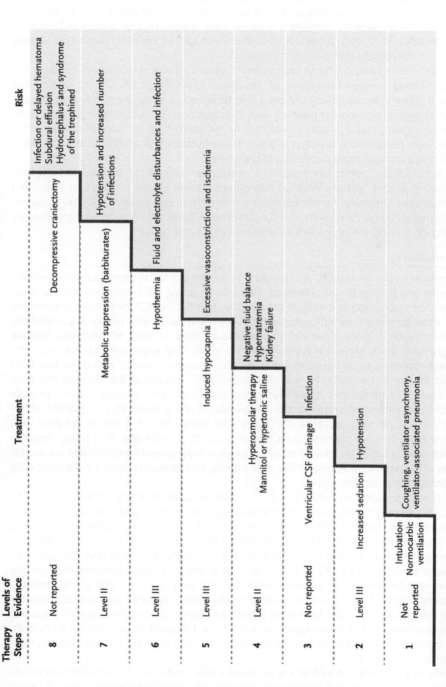

Fig. 4. Tiered approach to treatment of increased intracranial pressure. (*From* Stocchetti N, Maas AIR. Traumatic intracranial hypertension. N Engl J Med 2014;370(22):2126; with permission.)

on the clinical examination and CT imaging. Significant controversy followed the publication of this trial, which included a debate as to the value of ICP monitoring and the ability to generalize to developed countries. In response, a group of experts, including R.M. Chesnut, recently published a consensus statement that addressed many of these concerns.[49] Most experts agreed that the results of the study should not change practice for centers that already monitor ICP.

Brain tissue oxygen (Pbto$_2$) monitoring is used in many neurotrauma centers. The Brain Tissue Oxygen Monitoring in Traumatic Brain Injury (BOOST-2) clinical trial was a phase 2 randomized controlled trial designed to show the safety and efficacy of a treatment protocol for patients with TBI based on Pbto$_2$ monitoring. The investigators showed that a protocol designed to maintain Pbto$_2$ at more than 20 mm Hg effectively reduced the duration of brain tissue hypoxia.[50] A phase 3 trial to evaluate its impact on neurologic outcomes is underway.

Multimodality monitoring measures surrogate markers of cellular metabolism, including probes that are inserted into cerebral white matter to measure Pbto$_2$ and temperature, and cerebral microdialysis catheters that can measure lactate, pyruvate, glutamate, and glucose levels. Cerebral microdialysis is still considered experimental, although it is increasingly used at tertiary care trauma centers.

Osmotherapy

When sedation, intubation, and postural repositioning fail to decrease ICP, osmotic agents are frequently administered. Hypertonic saline and mannitol are the most commonly used agents and both are effective in reducing ICP. Both work presumably by establishing an osmotic gradient between the brain and cerebral vasculature, which results in a net water loss in brain tissue. Thus, in order to effectively achieve a gradient, the blood-brain barrier must be intact. The superiority of one particular agent continues to be debated, although until recently mannitol was considered the gold standard. Recently, Boone and colleagues[51] reviewed the existing literature comparing mannitol and hypertonic saline and found only a small number of trials with generally limited enrollment. A meta-analysis of the same studies found that hypertonic saline may decrease ICP more effectively and for a longer duration than mannitol.[52] Without definitive data, the decision to choose one drug rather than the other may be based on several factors, including the physical condition and comorbidities of the patient, drug availability, side effects, and clinician comfort in administering a potentially unfamiliar agent.

Hypertonic saline is administered in a variety of concentrations ranging from 2% to 23.4% and may be the ideal hyperosmolar agent in patients who also require volume resuscitation. Typically, 23.4% NaCl is administered as a bolus over 10 minutes, whereas less concentrated forms (3% NaCl) can be given either as a bolus or infusion. For patients whose initial serum Na level is 140 mmol/L, the initial goal may be to increase the Na level to 145 to 150 mmol/L. A stepwise, titrated increase in serum Na level may be needed for ongoing control of ICP, with an upper limit of 160 mmol/L. Care should be taken to avoid inducing heart failure in patients who would otherwise not tolerate a large fluid load or in those with hyponatremia, because of concerns about central pontine myelinolysis (which is rare). Frequent serum sodium and osmolarity checks are indicated. Solutions greater than 3% must be administered via central venous access.

Mannitol is dosed at 0.25 to 1 g/kg every 4 to 6 hours. Because mannitol is an osmotic diuretic, care should be taken in patients presumed to be hypovolemic. A serum osmolarity greater than 320 mOsm/L is often cited as the upper limit for mannitol administration because of concerns for inducing kidney injury. Although the theoretic concern exists, there is no definitive human evidence to support this claim. Instead, there may be a ceiling effect, above which increasing serum osmolarity has a limited effect on decreasing ICP. Mannitol may be redosed by calculating the osmolar gap (typically when <10 mOsm/L). Both hypertonic saline and mannitol should be weaned slowly, because the brain responds to the osmotic gradient by creating idiogenic osmoles, which could potentially reverse the osmotic gradient on discontinuation.[53,54]

Hypothermia

The therapeutic potential of hypothermia in patients with TBI has been noted from the time of Hippocrates (460–377 BC), who stated that "a man will survive longer in winter than in summer, whatever be the part of the head in which the wound is situated."[55] Contemporary thinking followed that if hypothermia protocols could impart a mortality benefit and neuroprotection to patients after cardiac arrest,[56,57] then it might also provide a similar benefit to patients with TBI. In patients with TBI, hypothermia has been used both as a prophylactic neuroprotectant, in which it is thought to reduce blood-brain barrier dysfunction, levels of excitatory neurotransmitters, free radical production, the cerebral metabolic rate, and inflammatory responses,[58] as well as a means to treat increased ICP by reducing cerebral blood flow and blood volume. Adverse effects of hypothermia include a higher incidence of cardiac arrhythmias, hypotension, pulmonary infections, coagulopathies, electrolyte derangements, hyperglycemia, and hypothermia-induced diuresis.[59–64] Until very recently, there has been conflicting evidence for and against the use of hypothermia in patients with TBI.

The 2007 BTF guidelines reported that there was insufficient evidence to support the routine use of prophylactic hypothermia and recommended further study.[47] These recommendations were based on a meta-analysis performed by the BTF using level III data demonstrating that hypothermia in TBI patients had no consistent or statistically significant mortality benefit. It was, however, unclear whether prophylactic hypothermia may have higher chances of reducing mortality when cooling was maintained for more than 48 hours. Despite a lack of mortality benefit, their analysis did indicate that therapeutic hypothermia was associated with significantly more favorable neurologic outcomes, defined by Glasgow Outcome Scale (GOS) scores of 4 or 5, compared to normothermic controls.

Recently, Andrews and colleagues[65] conducted The European Study of Therapeutic Hypothermia (32°C–35°C) for Intracranial Pressure Reduction after Traumatic Brain Injury (Eurotherm3235) trial. This international, multicenter, randomized controlled trial assessed the use of at least 48 hours of hypothermia to reduce increased ICP in patients with TBI. The trial aimed to enroll 600 patients; however, it was halted during interim analysis after 387 patients were enrolled because of safety concerns. In patients with TBI, therapeutic hypothermia plus standard care successfully reduced ICP but did not improve neurologic recovery compared with standard therapy. A favorable outcome with Extended GOS (GOS-E) scores of 5 to 8 occurred in 26% of patients in the hypothermia group and in 37% of patients in the control group. The Eurotherm3235 trial provides convincing evidence against the routine use of hypothermia as a means of reducing increased ICP after TBI.

Decompressive Craniectomy

Decompressive craniectomy is a treatment option for severe refractory intracranial hypertension, in which a substantial portion of the cranium is removed and the dura opened to increase the volume of the cranial cavity and, thus, decrease the ICP. This option is an effective means of controlling increased ICP; however, risks include wound infection, meningitis, cerebral abscess, CSF leak, hematoma, and cerebral infarction. There is divided neurosurgical opinion on the benefits of decompressive craniectomy in patients with TBI.

Although the 2007 BTF "Guidelines for the Surgical Management of Traumatic Brain Injury"[66] provide recommendations for surgical evacuation of posttraumatic intracranial mass lesions, there are no recommendations on the use of decompressive craniectomy for controlling increased ICP after severe TBI because of a lack of conclusive evidence of improved quality of life.

The Decompressive Craniectomy (DECRA) trial, published in 2011, was a randomized controlled trial that challenged the presumed benefits of the decompressive craniectomy in patients with TBI with intracranial hypertension. Patients with diffuse brain injury with intracranial hypertension refractory to first-tier therapies were randomized to bifrontotemporal decompressive craniectomy or standard care. Patients in the craniectomy group had less time with ICP above the treatment threshold, fewer interventions for increased ICP, and fewer days in the ICU compared with the control group. However, 6-month mortality was not affected, and patients who underwent craniectomy had worse outcome scores compared with the control group (odds ratio [OR] = 1.84; 95% confidence interval, 1.05–3.24; P = .03) and greater risk of an unfavorable outcome (OR = 2.21; 95% confidence interval, 1.14–4.26; P = .02). One explanation may be that craniectomy allows expansion of the swollen brain outside the skull but causes axonal stretch leading to neuronal injury.[67]

There have been many criticisms of this study. Baseline characteristics, including pupil reactivity, were imbalanced between groups. There was a high crossover rate from the standard-care arm to the surgical arm. The results have a poor ability to be generalized given that the study only assessed patients with diffuse brain injury less than the age of 60 years. Among other criticisms, the decision to use only bifrontotemporal craniectomy, which removes a large portion of the cranium, was seen as not following common clinical practice. Because of such criticisms, many authorities view these results with skepticism.

Randomized Evaluation of Intracranial Pressure (RESCUEicp) is an ongoing trial that compares medical therapy with decompressive craniectomy in patients with a sustained increase in ICP greater than 25 mm Hg resistant to initial medical therapy. Patients are randomized to continued medical therapy (including barbiturate use) or decompressive craniectomy. The results of this study will further define the role of decompressive craniectomy in patients with TBI.[14,44,68]

Steroids

Previously, steroid use in patients with TBI was controversial; however, at present the only BTF guideline that is classified as a level I recommendation is avoiding use of steroids in patients with moderate or severe TBI. This definitive recommendation is a result of the Corticosteroid Randomisation After Significant Head Injury (CRASH) multicenter trial published in 2004.[69] This study, which was designed to randomize 20,000 patients with TBI to 48 hours of intravenous methylprednisolone or placebo, was halted after interim analysis by the data monitoring committee after 10,008 patients were enrolled and analyzed. The interim analysis showed a significant difference in 2-week mortality, 21.1% in the steroid arm versus 17.9% in the placebo arm, and 6-month mortality, 25.7% in the steroid arm versus 22.3% in the placebo arm.[69,70]

Tranexamic Acid

The Clinical Randomization of an Antifibrinolytic in Significant Haemorrhage (CRASH-2) trial, published in 2010, assessed the effects of early administration of tranexamic acid to adult patients with trauma with or at risk of significant hemorrhage within

8 hours of injury. The trial showed significant reduction of all-cause mortality (Relative Risk = 0.91; 95% confidence interval, 0.85–0.97; P = .0035) with no increase in vascular occlusive events. Although there was no direct evidence, the investigators considered part of the mortality benefit of tranexamic acid to be a result of decreased intracranial bleeding in patients with trauma with brain injury.[71,72]

OUTCOMES

Moderate and severe TBI are associated with high mortality in the acute period. While managing the acute needs of patients with TBI, it is important to recognize the long-term morbidities of this condition. Management efforts must focus not only on mortality benefit but also on functional, neurocognitive, and emotional outcomes. There is increasing emphasis on studying outcomes beyond the acute phase of TBI. Various outcome scores have been developed to track and compare outcomes in patients with TBI, including the GOS and the Modified Rankin Scale (mRS).

Glasgow Outcome Scale and Extended Glasgow Outcome Scale

After moderate and severe TBI, there often is persisting mental and physical disability. As the apropos description indicates, the silent epidemic renders a patient after TBI with disabilities, especially mental disabilities that are difficult to detect.

In 1975, Jennett and Bond[73] published the 5-point GOS, which provided an objective, standardized, accurate assessment of outcome after severe brain damage. The GOS is the most widely used outcome measure after traumatic brain damage.[74] This scale allows monitoring of clinical progress to ensure adequate and efficacious care and rehabilitation. Many research studies use the GOS at 6 months after injury as a primary outcome measure because most of the improvement occurs during this period.[24] The 5-point scale is outlined in **Table 3**.

Table 3 GOS and GOS-E				
Score	GOS Grade	Explanation	Score	GOS-E Grade
1	Death	—	1	D
2	Persistent vegetative state	Patient shows no cortical function and remains unresponsive and speechless	2	VS
3	Severe disability	Patient is conscious but disabled and dependent on others for daily support because of mental or physical disability or both	3 4	SD− SD+
4	Moderate disability	Patient is disabled but independent in activities of daily living. Disabilities include varying degrees of dysphasia, hemiparesis, or ataxia as well as intellectual and memory deficits and personality change	5 6	MD− MD+
5	Good recovery	Patient has resumption of normal life even though there may be minor neurologic and psychological deficits	7 8	GR− GR+

Abbreviations: D, death; GR−, lower good recovery; GR+, upper good recovery; MD−, lower moderate disability; MD+, upper moderate disability; SD−, lower severe disability; SD+, upper severe disability; VS, vegetative state.

The creators of this scale hoped to prevent underestimation or overestimation of a patient's disabilities, and carefully decided on the definitions of each level. For example, good recovery was not classified as return to work because some patients who are generally functional may not be able to return to work because of socioeconomic factors, whereas others who may have considerable disability may be fully employed because their disability continues to be compatible with their work.

One of the criticisms of the GOS is that the categories are broad and insensitive to change, which prompted the development of the GOS-E. The GOS-E extends the original 5 GOS categories to 8, in which the original scale's highest 3 categories, 3 to 5, which describe patients who regain consciousness, are subdivided to allow more sensitivity in measurement of recovery. These 3 categories were subdivided into upper and lower divisions; however, no criteria were given for making these distinctions. Compared with the GOS, the GOS-E is more sensitive to change in mild to moderate TBI.[74,75]

The main criticisms of both the GOS and GOS-E are the reliability of the interviewer and the subjectivity involved in assigning the patient's score. Because of this, structured interviews have been created and are recommended to facilitate consistency in ratings.[74] Good interrater reliability and content validity have been shown for the GOS-E structured interview.[75]

Modified Rankin Scale

The original Rankin scale was introduced in 1957 by Dr John Rankin of Glasgow, United Kingdom.[76] The mRS was introduced by the Warlow group in the late 1980s to accommodate language disorders and cognitive defects.[77] The mRS has historically been used to assess recovery after stroke. However, the functionality it assesses is not specific to stroke, and thus it is widely used to assess neurologic dysfunction in a broad range of neurologic and neurosurgical conditions.[78] It is a 6-level ordinal outcome scale (0–5) used to assess the functional status of patients, encoding the range from no symptoms or functional impairment (mRS = 0) to severe disability requiring constant nursing care (mRS = 5) (**Table 4**). Unlike other scales, the mRS takes into account independence rather than performance of specific tasks, and in this way incorporates mental as well as physical adaptation to the neurologic deficits that patients face.

The first study to assess the interrater reliability of the mRS found it to be satisfactory[79]; however, there has been continued controversy over the subjective nature of

Table 4 mRS	
Score	Description
0	No symptoms
1	No significant disability despite symptoms; able to perform all usual duties and activities
2	Slight disability; unable to perform all previous activities, but able to look after own affairs without assistance
3	Moderate disability; requiring some help, but able to walk without assistance
4	Moderately severe disability; unable to walk without assistance and unable to attend to own bodily needs without assistance
5	Severe disability; bedridden, incontinent, requiring constant nursing care and attention

the scale. There have been multiple efforts to reduce the subjectivity and improve the interrater reliability of the mRS. These efforts include the creation of a structured interview that can be answered by the patient or caregiver over the phone or in person,[80] and multimedia training for the clinical personnel conducting the interview; Web-based calculators have also been created.[78]

Outcomes After Publication of the Brain Trauma Foundation Guidelines

The *Journal of Neurotrauma* has published 3 editions of the "Guidelines for the Management of Severe Traumatic Brain Injury" under the sponsorship of the BTF. The first was published in 1995, revised in 2000, and the most recent in 2007. There are only a few studies that examine the impact of application of these guidelines on TBI outcomes, and data show that compliance reduces mortality and improves outcomes.[81] Gerber and colleagues[82] calculated that TBI mortality between 2001 and 2009 in New York State experienced significant reduction from 22% to 13%, whereas there was a significant increase in guideline adherence, from 56% to 75%.

Arabi and colleagues[46] conducted a retrospective pre-post study in a single trauma center in Saudi Arabia and found a reduction in hospital mortality (OR = 0.45; 95% confidence interval, 0.24–0.86; P = .02) in patients with severe TBI after implementation of a management protocol that was derived from the 1995 and 2000 BTF guidelines. They also found that the use of the protocol was not associated with an increase in the need for tracheostomies, mechanical ventilation duration, or ICU or hospital length of stay, suggesting that the improved survival was not associated with an increased number of surviving patients with severe disability and that functional status might have also improved.

Faul and colleagues[2] applied a cost-benefit analysis of adoption and compliance of the 2000 BTF guidelines for the treatment of adults with severe TBI. They found that widespread adoption of these guidelines would result in substantial medical, rehabilitation, and societal savings in costs, with savings of $4.08 billion annually after accounting for implementation costs. They estimated that there would be a simultaneous mortality benefit and that 3607 lives would be saved annually. They also determined that the proportion of patients with good outcomes (GOS score 4–5) would increase from 35% to 66% and the proportion of patients with poor outcomes (GOS 2–3) would decrease from 34% to 19%, and thus the burden on families would likely be reduced.

REFERENCES

1. Faul M, Xu L, Wald M, et al. Traumatic brain injury in the United States: emergency department visits, hospitalizations and deaths 2002-2006. Atlanta (GA): Centers for Disease Control and Prevention; National Center for Injury Prevention and Control; 2010.

2. Faul M, Wald MM, Rutland-Brown W, et al. Using a cost-benefit analysis to estimate outcomes of a clinical treatment guideline: testing the Brain Trauma Foundation guidelines for the treatment of severe traumatic brain injury. J Trauma 2007;63(6):1271-8.

3. Selassie AW, Zaloshnja E, Langlois JA, et al. Incidence of long-term disability following traumatic brain injury hospitalization, United States, 2003. J Head Trauma Rehabil 2008;23(2):123-31.

4. Finkelstein EA, Corso PS, Miller TR. The incidence and economic burden of injuries in the United States. New York: Oxford University Press; 2006.

5. Slewa-Younan S, Green AM, Baguley IJ, et al. Sex differences in injury severity and outcome measures after traumatic brain injury. Arch Phys Med Rehabil 2004;85(3):376–9.
6. Cipolla M. The cerebral circulation. San Rafael (CA): Morgan & Claypool Life Sciences; 2009.
7. Czosnyka M, Smielewski P, Kirkpatrick P, et al. Monitoring of cerebral autoregulation in head-injured patients. Stroke 1996;27(10):1829–34.
8. Stiefel MF, Tomita Y, Marmarou A. Secondary ischemia impairing the restoration of ion homeostasis following traumatic brain injury. J Neurosurg 2005;103(4):707–14.
9. Bouma GJ, Muizelaar JP, Choi SC, et al. Cerebral circulation and metabolism after severe traumatic brain injury: the elusive role of ischemia. J Neurosurg 1991;75(5):685–93.
10. Bouma GJ, Muizelaar JP, Stringer WA, et al. Ultra-early evaluation of regional cerebral blood flow in severely head-injured patients using xenon-enhanced computerized tomography. J Neurosurg 1992;77(3):360–8.
11. Marion DW, Darby J, Yonas H. Acute regional cerebral blood flow changes caused by severe head injuries. J Neurosurg 1991;74(3):407–14.
12. Reilly PL. Brain injury: the pathophysiology of the first hours. 'Talk and Die revisited'. J Clin Neurosci 2001;8(5):398–403.
13. Werner C, Engelhard K. Pathophysiology of traumatic brain injury. Br J Anaesth 2007;99(1):4–9.
14. Hutchinson P, Corteen E, Czosnyka M, et al. Decompressive craniectomy in traumatic brain injury: the randomized multicenter RESCUEicp study (www. RESCUEicp. com). In: Hoff JT, Keep RF, Guohua Xi, et al, editors. Brain edema XIII. Austria: Springer; 2006. p. 17–20.
15. Moppett I. Traumatic brain injury: assessment, resuscitation and early management. Br J Anaesth 2007;99(1):18–31.
16. Chesnut RM, Marshall LF, Klauber MR, et al. The role of secondary brain injury in determining outcome from severe head injury. J Trauma 1993;34(2):216–22.
17. Fearnside MR, Cook RJ, McDougall P, et al. The Westmead Head Injury Project outcome in severe head injury. A comparative analysis of pre-hospital, clinical and CT variables. Br J Neurosurg 1993;7(3):267–79.
18. Hsiao AK, Michelson SP, Hedges JR. Emergent intubation and CT scan pathology of blunt trauma patients with Glasgow Coma Scale scores of 3–13. Prehosp Disaster Med 1993;8(03):229–36.
19. Teasdale G, Jennett B. Assessment of coma and impaired consciousness: a practical scale. Lancet 1974;304(7872):81–4.
20. Gill MR, Reiley DG, Green SM. Interrater reliability of Glasgow Coma Scale scores in the emergency department. Ann Emerg Med 2004;43(2):215–23.
21. Menegazzi JJ, Davis EA, Sucov AN, et al. Reliability of the Glasgow Coma Scale when used by emergency physicians and paramedics. J Trauma 1993;34(1):46–8.
22. Juarez VJ, Lyons M. Interrater reliability of the Glasgow Coma Scale. J Neurosci Nurs 1995;27(5):283–6.
23. Marmarou A, Lu J, Butcher I, et al. Prognostic value of the Glasgow Coma Scale and pupil reactivity in traumatic brain injury assessed pre-hospital and on enrollment: an IMPACT analysis. J Neurotrauma 2007;24(2):270–80.
24. Ghajar J. Traumatic brain injury. Lancet 2000;356(9233):923–9.
25. Sternbach GL. The Glasgow coma scale. J Emerg Med 2000;19(1):67–71.
26. Adoni A, McNett M. The pupillary response in traumatic brain injury: a guide for trauma nurses. J Trauma Nurs 2007;14(4):191–6.

27. Heiden JS, Small R, Caton W, et al. Severe head injury clinical assessment and outcome. Phys Ther 1983;63(12):1946–51.
28. A multicenter trial of the efficacy of nimodipine on outcome after severe head injury. The European Study Group on Nimodipine in Severe Head Injury. J Neurosurg 1994;80:797–804.
29. Jennett B, Teasdale G, Braakman R, et al. Predicting outcome in individual patients after severe head injury. Lancet 1976;307(7968):1031–4.
30. Andrews BT, Levy ML, Pitts LH. Implications of systemic hypotension for the neurological examination in patients with severe head injury. Surg Neurol 1987; 28(6):419–22.
31. Braakman R, Gelpke G, Habbema J, et al. Systematic selection of prognostic features in patients with severe head injury. Neurosurgery 1980;6(4):362–70.
32. van Dongen KJ, Braakman R, Gelpke GJ. The prognostic value of computerized tomography in comatose head-injured patients. J Neurosurg 1983;59(6):951–7.
33. Larner A. False localising signs. J Neurol Neurosurg Psychiatry 2003;74(4): 415–8.
34. Ritter AM, Muizelaar JP, Barnes T, et al. Brain stem blood flow, pupillary response, and outcome in patients with severe head injuries. Neurosurgery 1999;44(5): 941–8.
35. Chen JW, Gombart ZJ, Rogers S, et al. Pupillary reactivity as an early indicator of increased intracranial pressure: the introduction of the neurological pupil index. Surg Neurol Int 2011;2:82.
36. Hoffmann M, Lefering R, Rueger J, et al. Pupil evaluation in addition to Glasgow Coma Scale components in prediction of traumatic brain injury and mortality. Br J Surg 2012;99(S1):122–30.
37. Murray GD, Butcher I, McHugh GS, et al. Multivariable prognostic analysis in traumatic brain injury: results from the IMPACT study. J Neurotrauma 2007; 24(2):329–37.
38. Wilson SF, Amling JK, Floyd SD, et al. Determining interrater reliability of nurses' assessments of pupillary size and reaction. J Neurosci Nurs 1988;20(3):189–92.
39. Du R, Meeker M, Bacchetti P, et al. Evaluation of the portable infrared pupillometer. Neurosurgery 2005;57(1):198–203.
40. Taylor WR, Chen JW, Meltzer H, et al. Quantitative pupillometry, a new technology: normative data and preliminary observations in patients with acute head injury. Technical note. J Neurosurg 2003;98(1):205–13.
41. Marshall LF, Marshall SB, Klauber MR, et al. A new classification of head injury based on computerized tomography. Spec Supplements 1991;75(1S):S14–20.
42. Marshall L, Marshall SB, Klauber M, et al. The diagnosis of head injury requires a classification based on computed axial tomography. J Neurotrauma 1992;9: S287–92.
43. Gilbert K, Havill J, Sleigh J, et al. Observer error and prediction of outcome-grading of head injury based on computerised tomography. Crit Care Resusc 2001;3(1):15–8.
44. Stocchetti N, Maas AI. Traumatic intracranial hypertension. N Engl J Med 2014; 370(22):2121–30.
45. Ghajar J, Hariri RJ, Narayan RK, et al. Survey of critical care management of comatose, head-injured patients in the United States. Crit Care Med 1995; 23(3):560–7.
46. Arabi YM, Haddad S, Tamim HM, et al. Mortality reduction after implementing a clinical practice guidelines–based management protocol for severe traumatic brain injury. J Crit Care 2010;25(2):190–5.

47. Brain Trauma Foundation, American Association of Neurological Surgeons, Congress of Neurological Surgeons. Guidelines for the management of severe traumatic brain injury. J Neurotrauma 2007;24(Suppl 1):S1–106.
48. Chesnut RM, Temkin N, Carney N, et al. A trial of intracranial-pressure monitoring in traumatic brain injury. N Engl J Med 2012;367(26):2471–81.
49. Chesnut RM, Bleck TP, Citerio G, et al. A consensus-based interpretation of the benchmark evidence from South American trials: treatment of intracranial pressure trial. J Neurotrauma 2015;32(22):1722–4.
50. Brain tissue oxygen monitoring in traumatic brain injury (TBI) (BOOST 2). Available at: https://clinicaltrials.gov/ct2/show/NCT00974259. Accessed November 4, 2015.
51. Boone MD, Oren-Grinberg A, Robinson TM, et al. Mannitol or hypertonic saline in the setting of traumatic brain injury: what have we learned? Surg Neurol Int 2015; 6:177.
52. Li M, Chen T, Cai J, et al. Comparison of equimolar doses of mannitol and hypertonic saline for the treatment of elevated intracranial pressure after traumatic brain injury: a systematic review and meta-analysis. Medicine 2015;94(17):e736.
53. Rudehill A, Gordon E, Ohman G, et al. Pharmacokinetics and effects of mannitol on hemodynamics, blood and cerebrospinal fluid electrolytes, and osmolality during intracranial surgery. J Neurosurg Anesthesiol 1993;5(1):4–12.
54. Kofke AW. Mannitol: potential for rebound intracranial hypertension? J Neurosurg Anesthesiol 1993;5(1):1–3.
55. Adams F. The genuine works of Hippocrates. Baltimore (MD): Williams & Wilkins; 1939.
56. Bernard SA, Gray TW, Buist MD, et al. Treatment of comatose survivors of out-of-hospital cardiac arrest with induced hypothermia. N Engl J Med 2002;346(8):557–63.
57. Hypothermia after Cardiac Arrest Study Group. Mild therapeutic hypothermia to improve the neurologic outcome after cardiac arrest. N Engl J Med 2002;346(8):549–56.
58. Adamides AA, Winter CD, Lewis PM, et al. Current controversies in the management of patients with severe traumatic brain injury. ANZ J Surg 2006;76(3):163–74.
59. Clifton GL, Miller ER, Choi SC, et al. Lack of effect of induction of hypothermia after acute brain injury. N Engl J Med 2001;344(8):556–63.
60. Clifton G, Allen S, Berry J, et al. Systemic hypothermia in treatment of brain injury. J Neurotrauma 1992;9:S487–95.
61. Bernard SA, Jones BM, Buist M. Experience with prolonged induced hypothermia in severe head injury. Crit Care 1999;3(6):167.
62. Shiozaki T, Hayakata T, Taneda M, et al. A multicenter prospective randomized controlled trial of the efficacy of mild hypothermia for severely head injured patients with low intracranial pressure. J Neurosurg 2001;94(1):50–4.
63. Polderman KH, Peerdeman SM, Girbes AR. Hypophosphatemia and hypomagnesemia induced by cooling in patients with severe head injury. J Neurosurg 2001;94(5):697–705.
64. Clifton GL, Allen S, Barrodale P, et al. A phase II study of moderate hypothermia in severe brain injury. J Neurotrauma 1993;10(3):263–71.
65. Andrews PJ, Sinclair HL, Rodriguez A, et al. Hypothermia for intracranial hypertension after traumatic brain injury. N Engl J Med 2015;373(25):2403–12.
66. Bullock RM, Chesnut RM, Ghajar J, et al. Guidelines for the surgical management of traumatic brain injury. Neurosurgery 2006;58. S2-1–S2-62.

67. Cooper DJ, Rosenfeld JV, Murray L, et al. Decompressive craniectomy in diffuse traumatic brain injury. N Engl J Med 2011;364(16):1493–502.
68. Hutchinson PJ, Timofeev I, Kolias AG, et al. Decompressive craniectomy for traumatic brain injury: the jury is still out. Br J Neurosurg 2011;25(3):441–2.
69. Roberts I, Yates D, Sandercock P, et al. Effect of intravenous corticosteroids on death within 14 days in 10008 adults with clinically significant head injury (MRC CRASH trial): randomised placebo-controlled trial. Lancet 2004; 364(9442):1321–8.
70. Edwards P, Arango M, Balica L, et al. Final results of MRC CRASH, a randomised placebo-controlled trial of intravenous corticosteroid in adults with head injury- outcomes at 6 months. Lancet 2005;365(9475):1957–9.
71. CRASH-2 Collaborators, Roberts I, Shakur H, Afolabi A, et al. The importance of early treatment with tranexamic acid in bleeding trauma patients: an exploratory analysis of the CRASH-2 randomised controlled trial. Lancet 2011;377(9771): 1096–101.e1.
72. CRASH-2 Trial Collaborators, Shakur H, Roberts I, Bautista R, et al. Effects of tranexamic acid on death, vascular occlusive events, and blood transfusion in trauma patients with significant haemorrhage (CRASH-2): a randomised, placebo-controlled trial. Lancet 2010;376(9734):23–32.
73. Jennett B, Bond M. Assessment of outcome after severe brain damage: a practical scale. Lancet 1975;305(7905):480–4.
74. Wilson JL, Pettigrew LE, Teasdale GM. Structured interviews for the Glasgow Outcome Scale and the extended Glasgow Outcome Scale: guidelines for their use. J Neurotrauma 1998;15(8):573–85.
75. Sander A. The extended Glasgow Outcome Scale. The Center for Outcome Measurement in Brain Injury. 2002. Available at: http://www.tbims.org/combi/gose. Accessed October 1, 2015.
76. Rankin J. Cerebral vascular accidents in patients over the age of 60. II. Prognosis. Scott Med J 1957;2(5):200–15.
77. Farrell B, Godwin J, Richards S, et al. The United Kingdom Transient Ischaemic Attack (UK-TIA) aspirin trial: final results. J Neurol Neurosurg Psychiatry 1991; 54(12):1044–54.
78. Patel N, Rao VA, Heilman-Espinoza ER, et al. Simple and reliable determination of the modified Rankin Scale score in neurosurgical and neurological patients: the mRS-9Q. Neurosurgery 2012;71(5):971–5 [discussion: 975].
79. Van Swieten J, Koudstaal P, Visser M, et al. Interobserver agreement for the assessment of handicap in stroke patients. Stroke 1988;19(5):604–7.
80. Wilson JL, Hareendran A, Grant M, et al. Improving the assessment of outcomes in stroke use of a structured interview to assign grades on the Modified Rankin Scale. Stroke 2002;33(9):2243–6.
81. Watts DD, Hanfling D, Waller MA, et al. An evaluation of the use of guidelines in prehospital management of brain injury. Prehosp Emerg Care 2004;8(3):254–61.
82. Gerber LM, Chiu Y-L, Carney N, et al. Marked reduction in mortality in patients with severe traumatic brain injury: clinical article. J Neurosurg 2013;119(6): 1583–90.

Subarachnoid Hemorrhage
An Update

Jeremy S. Dority, MD*, Jeffrey S. Oldham, MD

KEYWORDS

- Subarachnoid hemorrhage • Delayed cerebral ischemia • Vasospasm
- Cerebral salt wasting • Coiling

KEY POINTS

- Target euvolemia, not hypervolemia, in the management of delayed cerebral ischemia in subarachnoid hemorrhage.
- Delayed cerebral ischemia is the most prominent complication of subarachnoid hemorrhage and can occur independent of angiographic vasospasm.
- Based on large, randomized trials, initiation of statin therapy does not improve outcomes in subarachnoid hemorrhage.
- Oral nimodipine therapy remains the mainstay of neuroprotective therapy in subarachnoid hemorrhage.
- Hyponatremia is common in subarachnoid hemorrhage and is associated with longer length of stay, but not increased mortality.

INTRODUCTION

Subarachnoid hemorrhage (SAH) is a debilitating, although uncommon, type of stroke with high morbidity, mortality, and economic impact. Occurring with an incidence between 2 and 22 per 100,000 persons per year with regional world variation, SAH accounts for about 5% of all strokes. In the United States, the incidence is about 10 per 100,000 persons per year.[1] Modern 30-day mortality is as high as 40%, and about 50% of survivors have permanent disability. Care at high-volume centers with dedicated neurointensive care units is recommended, although subspecialty expertise may be more important than clinical volume. Euvolemia, not hypervolemia, should be targeted, and the aneurysm should be secured early. Although nimodipine remains the mainstay of treatment for neurologic protection after SAH, neither statin nor

This work is published in collaboration with the Society for Neuroscience in Anesthesiology and Critical Care.

Disclosure: The authors have nothing to disclose.

Department of Anesthesiology, University of Kentucky College of Medicine, 800 Rose Street, Suite N202, Lexington, KY 40536-0293, USA

* Corresponding author.

E-mail address: jsdori2@uky.edu

Anesthesiology Clin 34 (2016) 577–600

http://dx.doi.org/10.1016/j.anclin.2016.04.009   anesthesiology.theclinics.com magnesium infusions should be initiated for delayed cerebral ischemia (DCI). Cerebral vasospasm is just one component of DCI. Hyponatremia is common in SAH and is associated with longer length of stay, but not increased mortality. This article focuses on selected points of management of these critically ill patients with an emphasis on the newest understanding and care recommendations relevant to anesthesiologists and neurointensivists.

MANAGEMENT BEFORE ANEURYSM OBLITERATION

Anesthesiologists are typically consulted after diagnosis of SAH has been made. Evaluation of the patient's ability to protect the airway and hemodynamic monitoring and control are paramount in this first assessment, especially as the patient is transported to the intensive care unit or prepared for aneurysm obliteration in the angiography or operating suite (**Fig. 1**).

Neurologic examination, including mental status, is important, because anesthesia masks subtle changes. Ensuing neurologic deterioration should be expected and plans developed (**Table 1**).

An aneurysm that has caused an SAH can rebleed. Although the size of the aneurysm is the strongest predictor of rupture, rebleeding of the aneurysm may be partly attributable to uncontrolled hypertension. A titratable agent (eg, nicardipine) is preferentially used to prevent extreme hypertension, and specific blood pressure goals should be individualized based on the patient's age, cardiac and baseline blood pressure history, and aneurysm size. Hypotension should also be avoided because it may cause oligemia through compromised cerebral perfusion pressure and increase the risk and size of stroke. The Interpretation and Implementation of Intensive Blood Pressure Reduction in Acute Cerebral Hemorrhage Trial (INTERACT-2)[2] showed the safety of targeting systolic blood pressure less than 140 mm Hg compared with a target of less than 180 mm Hg. Similarly, The Intracerebral Haemorrhage Acutely Decreasing Arterial Pressure Trial (ICH ADAPT)[3] trial compared a

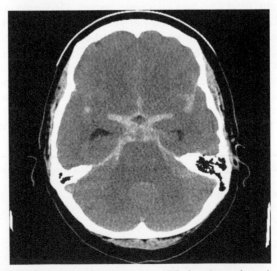

Fig. 1. Noncontrast head computed tomography (CT) showing subarachnoid blood. (*Courtesy of* Justin Fraser, MD, Lexington, KY.)

Table 1
Time course can be used to prioritize differential considerations of causes of acute neurologic decline in SAH

Early: First 24 h	Late: Days 3–10
Rebleed	Vasospasm
Acute hydrocephalus	Infection
Seizure	Intracranial hypertension
—	Delirium
—	DCI
—	Seizure
—	Sodium abnormalities
—	Pituitary dysfunction

target of less than 150 mm Hg with less than 180 mm Hg and did not find a reduction in blood flow on computed tomography (CT) perfusion. Our practice is to target systolic blood pressure less than 140 mm Hg using nicardipine as the first-line agent.[4] Antihypertensive Treatment of Acute Cerebral Hemorrhage (ATACH-II),[5] a multicenter, randomized, controlled, phase III trial, has enrolled 1000 subjects and will evaluate intensive systolic blood pressure reduction to 140 mm Hg using nicardipine compared with 180 mm Hg in the control arm. Ultimately, definitive treatment to prevent rebleeding involves aneurismal obliteration. The aneurysm should be obliterated through neurointerventional approaches or operative microsurgical clipping as early as feasible, because the risk of rehemorrhage is up to 26% in the first 2 weeks without treatment.[6] Early aneurysm obliteration is associated with reduced mortality and is likely the most significant improvement in care for patients with SAH.[7–9]

Antifibrinolytics: A Resurgence?

Historically, surgical intervention was delayed given the challenges of operating on an edematous, inflamed brain. To lessen the risk of early rebleeding, antifibrinolytic agents were administered. Because rebleeding was thought to be caused by breakdown of fresh thrombus on the aneurysm wall, tranexamic acid or ε-aminocaproic acid were administered to prevent clot lysis through plasminogen activation. However, this approach was associated with increased risk of cerebral ischemia and infarction, thromboembolism, and shunt-dependent hydrocephalus.[10] Subsequently, de Gans and Siddiq[7,9] showed that early (<24 or 48 hours) aneurysm occlusion improved outcomes, primarily through reducing risk of rebleeding. As such, current recommendations are to secure the aneurysm as early as feasible. In the event of an insurmountable delay in microsurgical or endovascular treatment, antifibrinolytic therapy remains an option to reduce the risk of rebleed (**Table 2**). Although this renewed but short-term application of antifibrinolytic drugs is promising, further studies are needed to definitively establish their roles in the care of these patients. It should be emphasized that antifibrinolytic agents offer a possible temporizing approach and should never be applied to significantly delay aneurysm obliteration.[11] A prospective, randomized, multicenter study evaluating the effect of ultraearly tranexamic acid after SAH is underway.[12]

Table 2
Conclusions and recommendations for management of SAH

Treatment Decision	American Heart Association/American Stroke Association	Neurocritical Care Society	European Stroke Organization Guidelines for Management of SAH
Hospital/system characteristics	Low-volume hospitals (eg, <10 SAH cases per year) should consider early transfer of patients with SAH to high-volume centers (eg, >35 SAH cases per year) with experienced cerebrovascular surgeons, endovascular specialists, and multidisciplinary neurointensive care services (class I, level B) After discharge, it is reasonable to refer patients with SAH for comprehensive evaluation, including cognitive, behavioral, and psychosocial assessments (class IIa, level B)	Patients with SAH should be treated at high-volume centers (moderate quality of evidence, strong recommendation) High-volume centers should have appropriate specialty neurointensive care units, neurointensivists, vascular neurosurgeons, and interventional neuroradiologists to provide the essential elements of care (moderate quality of evidence, strong recommendation)	—

| Aneurysm treatment | Surgical clipping or endovascular coiling of the ruptured aneurysm should be performed as early as feasible in most patients to reduce the rate of rebleeding after SAH (class I, level B)

For patients with ruptured aneurysms judged to be technically amenable to either endovascular coiling or neurosurgical clipping, endovascular coiling should be considered (class I, level B)

Complete obliteration of the aneurysm is recommended whenever possible (class I, level B)

Stenting of a ruptured aneurysm is associated with increased morbidity and mortality (class III, level C)

For patients with an unavoidable delay in obliteration of aneurysm, a significant risk of rebleeding, and no compelling medical contraindications, short-term (<72 h) therapy with tranexamic acid or aminocaproic acid is reasonable to reduce the risk of early aneurysm rebleeding (class IIa, level B) | Early aneurysm repair should be undertaken when possible and reasonable to prevent rebleeding (high quality of evidence, strong recommendation)

An early, short course of antifibrinolytic therapy before early aneurysm repair (begun at diagnosis and continued up to the point at which the aneurysm is secured or at 72 h postictus, whichever is shorter) should be considered (low quality of evidence, weak recommendation)

Delayed (>48 h after the ictus) or prolonged (>3 d) antifibrinolytic therapy exposes patients to side effects of therapy when the risk of rebleeding is sharply reduced and should be avoided (high quality of evidence, strong recommendation) | Intensive continuous observation at least until occlusion of the aneurysm

Continuous electrocardiogram monitoring

Start with GCS, focal deficits, blood pressure, and temperature at least every hour

Aneurysm should be treated as early as logistically and technically possible to reduce the risk of rebleeding; if possible it should be planned to intervene at least within 72 h after onset of first symptoms

This decision should not depend on grading (class III, level C)

The best mode of intervention should be discussed in an interdisciplinary dialogue between neurosurgery and neuroradiology

Based on this discussion, patients should be informed and included in the process of decision making whenever possible

If the aneurysm seems to be equally effectively treated either by coiling or clipping, coiling is the preferred treatment (class I, level A) |

(continued on next page)

Table 2
(continued)

Treatment Decision	American Heart Association/American Stroke Association	Neurocritical Care Society	European Stroke Organization Guidelines for Management of SAH
Blood pressure control	Between the time of SAH symptom onset and aneurysm obliteration, blood pressure should be controlled with a titratable agent to balance the risk of stroke, hypertension-related rebleeding, and maintenance of cerebral perfusion pressure (class I, level B) The magnitude of blood pressure control to reduce the risk of rebleeding has not been established, but a decrease in systolic blood pressure to <160 mm Hg is reasonable (class IIa, level C)	Treat extreme hypertension in patients with an unsecured, recently ruptured aneurysm. Modest increases in blood pressure (mean blood pressure of <110 mm Hg) do not require therapy. Premorbid baseline blood pressures should be used to refine targets and hypotension should be avoided (low quality of evidence, strong recommendation)	Stop antihypertensive medication that the patient was using Do not treat hypertension unless it is extreme; limits for extreme blood pressures should be set on an individual basis, taking into account age of the patient, pre-SAH blood pressures, and cardiac history; systolic blood pressure should be kept <180 mm Hg, only until coiling or clipping of ruptured aneurysm, to reduce risk for rebleeding If systolic pressure remains high despite these treatments, further lowering of blood pressure should be considered (class IV, level C) If the blood pressure is lowered the mean arterial pressure should be kept at least >90 mm Hg (GCP)
Intravascular volume status	Maintenance of euvolemia and normal circulating blood volume is recommended to prevent DCI (class I, level B)	Intravascular volume management should target euvolemia and avoid prophylactic hypervolemic therapy. In contrast, there is evidence for harm from aggressive administration of fluid designed to achieve hypervolemia (moderate quality of evidence, strong recommendation)	—

Cardiopulmonary complications	No recommendations given	Baseline cardiac assessment with serial enzymes, ECG, and echocardiography is recommended, especially in patients with evidence of myocardial dysfunction (low quality of evidence, strong recommendation) Monitoring of cardiac output may be useful in patients with evidence of hemodynamic instability or myocardial dysfunction (low quality of evidence, strong recommendation)	—
Seizures	The use of prophylactic anticonvulsants may be considered in the immediate posthemorrhagic period (class IIb, level B) The routine long-term use of anticonvulsants is not recommended (class III, level B)	Routine use of anticonvulsant prophylaxis with phenytoin is not recommended after SAH (low quality of evidence, strong recommendation) If anticonvulsant prophylaxis is used, a short course (3–7 d) is recommended (low quality of evidence, weak recommendation) Continuous electroencephalogram monitoring should be considered in patients with poor-grade SAH who fail to improve or who have neurologic deterioration of undetermined cause (low quality of evidence, strong recommendation)	Antiepileptic treatment should be administered in patients with clinically apparent seizures (GCP) There is no evidence that supports the prophylactic use of antiepileptic drugs (class IV, level C)

(continued on next page)

Table 2
(continued)

Treatment Decision	American Heart Association/American Stroke Association	Neurocritical Care Society	European Stroke Organization Guidelines for Management of SAH
Fever treatment	Aggressive control of fever to a target of normothermia by use of standard or advanced temperature-modulating systems is reasonable in the acute phase of SAH (class IIa, level B)	During the period of risk for DCI, control of fever is desirable; intensity should reflect the individual patient's relative risk of ischemia (low quality of evidence, strong recommendation) Surface cooling or intravascular devices are more effective and should be used when antipyretics fail in cases in which fever control is highly desirable (high quality of evidence, strong recommendation)	There are neither controlled studies on the effect of cooling in patients with SAH nor studies that have shown that treatment of fever does improve outcome Increased temperature should be treated medically and physically (GCP)
Glucose control	Careful glucose management with strict avoidance of hypoglycemia may be considered as part of the general critical care management of patients with SAH (class IIb, level B)	Hypoglycemia (serum glucose level of <80 mg/dL) should be avoided (high quality of evidence, strong recommendation) Serum glucose level should be maintained at <200 mg/dL (moderate quality of evidence, strong recommendation)	Hyperglycemia >10 mmol/L (180 mg/dL) should be treated (GCP)
Deep venous thrombosis prophylaxis	Heparin-induced thrombocytopenia and deep venous thrombosis are frequent complications after SAH. Early identification and targeted treatment are recommended, but further research is needed to identify the ideal screening paradigms (class I, level B)	Measures to Prevent deep venous thrombosis should be used in all patients with SAH (high quality of evidence, strong recommendation) The use of unfractionated heparin for prophylaxis could be started 24 h after undergoing aneurysm obliteration (moderate quality of evidence, strong recommendation)	Patients with SAH may be given thromboprophylaxis with pneumatic devices and/or compression stockings before occlusion of the aneurysm (class II, level B) In case deep venous thrombosis prevention is indicated, low-molecular-weight heparin should be applied not earlier than 12 h after surgical occlusion of the aneurysm and immediately after coiling (class II, level B) Compression stockings and intermittent compression by pneumatic devices in high-risk patients

DCI	Oral nimodipine should be administered to all patients with SAH (class I, level A)	Oral nimodipine (60 mg every 4 h) should be administered after SAH for a period of 21 d (high quality of evidence, strong recommendation)	Nimodipine should be administered orally (60 mg/4 h) to prevent delayed ischemic events (class I, level A)
	Maintenance of euvolemia and normal circulating blood volume is recommended to prevent DCI (class I, level B)	The goal should be maintaining euvolemia, rather than attempting hypervolemia (moderate quality of evidence, strong recommendation)	If oral administration is not possible nimodipine should be applied intravenously (GCP)
	Prophylactic hypervolemia or balloon angioplasty before the development of angiographic spasm is not recommended (class III, level B)	Transcranial Doppler may be used for monitoring and detection of large artery vasospasm with variable sensitivity (moderate quality of evidence, strong recommendation)	Magnesium sulfate is not recommended for prevention of DCI (class I, level A) Statins are < study
	Transcranial Doppler is reasonable to monitor for the development of arterial vasospasm (class IIa, level B)	Digital subtraction angiography is the gold standard for detection of large artery vasospasm (high quality of evidence, strong recommendation)	There is no evidence from controlled studies for induced hypertension or hypervolemia to improve outcome in patients with delayed ischemic deficit (class IV, level C)
	Perfusion imaging with CT or MRI can be useful to identify regions of potential brain ischemia (class IIa, level B)	Patients clinically suspected of DCI should undergo a trial of induced hypertension (moderate quality of evidence, strong recommendation)	
	Induction of hypertension is recommended for patients with DCI unless blood pressure is increased at baseline or cardiac status precludes it (class I, level B)	Endovascular treatment using intra-arterial vasodilators and/or angioplasty may be considered for vasospasm-related DCI (moderate quality of evidence, strong recommendation)	
	Cerebral angioplasty and/or selective intra-arterial vasodilator therapy is reasonable in patients with symptomatic vasospasm, particularly those who are not responding to hypertensive therapy (class IIa, level B)		
Anemia and transfusion	The use of packed red blood cell transfusion to treat anemia might be reasonable in patients with SAH who are at risk of cerebral ischemia. The optimal hemoglobin goal is still to be determined (class IIb, level B)	Patients should receive packed red blood cell transfusions to maintain hemoglobin concentration >8–10 g/dL (moderate quality of evidence, strong recommendation)	—

(continued on next page)

Table 2
(continued)

Treatment Decision	American Heart Association/American Stroke Association	Neurocritical Care Society	European Stroke Organization Guidelines for Management of SAH
Hyponatremia	The use of fludrocortisone acetate and hypertonic saline solution is reasonable for preventing and correcting hyponatremia (class IIa, level B)	Fluid restriction should not be used to treat hyponatremia (weak quality of evidence, strong recommendation) Early treatment with hydrocortisone of fludrocortisone may be used to limit natriuresis and hyponatremia (moderate quality of evidence, weak recommendation) Mild hypertonic saline solutions can be used to correct hyponatremia (very low quality of evidence, strong recommendation)	There is no proof that steroids are effective in patients with SAH (class IV, level C)
Hydrocephalus	—	—	In patients with CT-proven hydrocephalus and the third or fourth ventricle filled with blood, an external ventricular drain should be applied; this drain can be used to reduce and monitor pressure and to remove blood; for this last reason the level of evidence is low (GCP) In patients who are not seated and who deteriorate from acute hydrocephalus, lumbar puncture might be considered if the third and fourth ventricle are not filled with blood and supratentorial herniation is prevented (class IV, level C) In patients who are sedated and have CT-proven hydrocephalus, lumbar drainage should be considered if the third and fourth ventricles are not filled with blood (class IV, level C) Patients with symptomatic chronic hydrocephalus require ventriculoperitoneal or ventriculoatrial shunting (GCP)

Abbreviations: GCP, good clinical practice; GCS, Glasgow Coma Scale.

ANEURYSM OBLITERATION
Endovascular Evolution

Although various factors may make surgical clipping versus endovascular embolization preferable, a clear consensus is not yet held for the individualized care of most patients[13–15] (**Table 3**). Surgical clipping offers the advantage of immediate aneurysm exclusion from the cerebral circulation and direct irrigation of blood burden, which is associated with cerebral vasospasm. It is also associated with a lower risk of aneurysm recurrence or rebleed.[13] However, the patient is exposed to an open craniotomy with surgical brain retraction, and Molyneux and colleagues[16] in 2005 showed an increased rate of seizures (4.1% vs 2.5%). The primary benefit of endovascular techniques (summarized later), is the less invasive nature of the procedure. Understanding of the rapid technical progress of endovascular techniques is necessary for fruitful conversations with proceduralists and warrants a brief summary here.

COILING

Considerable progress has been made in endovascular approaches to cerebral aneurysm obliteration. Beginning with Guglielmi and colleagues[17] in 1991 with the first detachable coils, many patients have been treated successfully with variations on this minimally invasive approach[13,14] (**Fig. 2**).

Experience with this technique revealed necessary anatomic considerations. For example, Standhardt and colleagues[18] found in 2008 that the successful aneurysmal occlusion rate was considerably lower in wide-neck aneurysms (dome/neck ratio <2 or neck size ≥4 mm): 77.1% versus 35.8%. Another corollary of the shape of the aneurysm and its neck is the concern that coils will protrude into the parent arterial lumen, possibly occluding flow and causing thromboembolic complications. In an effort to mitigate this risk, the remodeling or balloon-assisted coiling technique was developed.[19]

Table 3		
Factors affecting approach to aneurysm obliteration		
Factor	**Angiographic Intervention (Coiling)**	**Operative Intervention (Clipping)**
Aneurysm properties	• Posterior circulation aneurysms • Basilar tip aneurysms • Intercavernous internal carotid artery aneurysms • Small aneurysm neck • Top-of-the-basilar aneurysm	• Middle cerebral artery aneurysms • Fusiform aneurysms • Aneurysms with wide neck • Aneurysms at arterial bifurcations • Aneurysm associated with large parenchymal hematoma • Normal arterial branches arising from dome or body of aneurysm
Age	≥70 y	Younger patients
Intracerebral hemorrhage properties	Space-occupying ICH not present	Presence of space-occupying ICH
Clinical considerations	• Comorbidities • Poor clinical grade • High surgical risk • Clinical stasis	—

Abbreviation: ICH, intracranial hemorrhage.

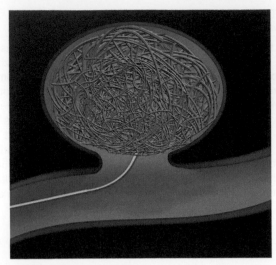

Fig. 2. Saccular aneurysm with coils in place.

Balloon-assisted Coiling

In Pierot and colleagues'[20] study (Aneurysms treated by endovascular approach Trial 2008), 37% of 739 nonruptured aneurysms were treated using balloon-assisted coiling (**Fig. 3**). They saw a reduction from 7.3% to 5.5% thromboembolic complications using this technique, but a review from Pierot and colleagues[21] in 2012 showed an increased risk of procedural complications with balloon-assisted coiling of 14.1%, versus 3.0% seen with coiling without balloon assist.

Stent-assisted Coiling

With the development of the self-expanding stents made from nitinol, stent-assisted coiling (SAC) became an alternative to balloon-assisted coiling. Nitinol is a unique nickel titanium metal with exceptional shape memory and superelasticity, which allow it to automatically expand to the desired shape on deployment.[22] Piotin and colleagues[23] in 2010 retrospectively analyzed 206 SACs and, compared with 1109 treated with coiling only, found a lower rate of aneurysm recurrence (14.9% vs 33.5%, respectively). Higher long-term aneurysm occlusion rates have been found in multiple studies[24–26] (**Fig. 4**).

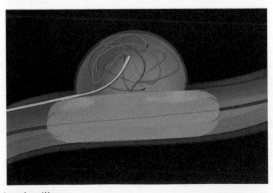

Fig. 3. Balloon-assisted coiling.

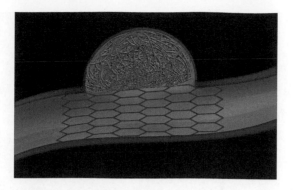

Fig. 4. SAC.

Flow Diversion

Further drawing on this evolution of experience led to the development of flow diverters. By using a higher mesh density than the stents used in SAC, the aneurysm can be obliterated over time by diverting flow through the parent artery instead of through the high-resistance mesh into the aneurysm. Only 1 such device, The Pipeline Embolization Device (Chestnut Medical Technologies, Menlo Park, CA), is approved for use in the United States.[27] Over time, blood stasis in the aneurysm and stent endothelialization lead to aneurysm obliteration. In the Pipeline Embolization Device for the Intracranial Treatment of Aneurysm (PITA) trial, 93.3% of treated aneurysms were fully occluded at 180-day follow-up.[28] Notably, this approach is not recommended for ruptured aneurysms, because it does not lead to rapid, complete obliteration and is associated with increased mortality in aneurysmal SAH. In the multicenter study by Lylyk and colleagues[29] in 2009, of 53 patients with 63 aneurysms, only 8% had immediate aneurysm occlusion (**Fig. 5**).

NEUROCRITICAL CARE
Neurogenic Pulmonary Edema

Neurogenic pulmonary edema is defined as any respiratory compromise that accompanies an acute neurologic insult that cannot be explained by coexisting cardiac or pulmonary derangement.[30] Although a nebulous definition, this disorder can complicate an already tenuous clinical course for patients following SAH. It is estimated that up to 23% of patients with SAH develop some form of pulmonary edema, with 2% to 8% being attributed to neurogenic pulmonary edema.[30,31] Of those patients who develop neurogenic pulmonary edema, mortality has been reported as high as 50%.[30]

The clinical presentation of neurogenic pulmonary edema includes dyspnea, tachypnea, tachycardia, cyanosis, pink frothy sputum, and/or crackles on auscultation of lung fields. Radiographic imaging reveals bilateral diffuse alveolar infiltrates[30] that may develop with early onset, beginning within minutes to hours, or delayed onset, beginning 12 to 24 hours after the neurologic insult.[30]

The pathophysiology behind neurogenic pulmonary edema is not fully understood but may be related to excess sympathetic nervous system activity. This sympathetic surge presumably causes pulmonary venoconstriction with a concomitant increase in capillary permeability from the release of inflammatory cytokines, which results in an increase in extravascular lung water and impaired oxygenation.[30,32] Also, lungs can incur direct damage from release of inflammatory cytokines triggered by either global or localized ischemia to areas of the hypothalamus and medulla oblongata, which

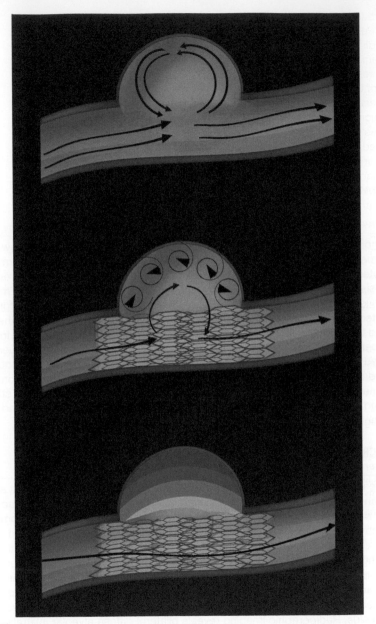

Fig. 5. Flow diversion.

causes leaky alveolar capillaries and extravasation of fluid into the lung parenchyma.[30] Pulmonary edema may be exacerbated by aggressive fluid resuscitation to treat vasospasm.

Treatment of neurogenic pulmonary edema is largely supportive. Although hypervolemia was previously endorsed in the management of SAH, lack of supportive outcome data has led to a current trend towards normovolemia.[31] Lung protective measures such as low tidal volume, administration of positive end-expiratory pressure and titration of inhaled oxygen fraction to maintain oxygen saturation are recommended.[30] Permissive hypercapnia, a lung protective technique often used in acute respiratory distress syndrome, should be avoided because this exacerbates increased intracranial pressure.

New research in the treatment of neurogenic pulmonary edema has targeted antagonism of the P2X7 receptor, an adenosine triphosphate–gated ion channel involved in cytotoxicity and apoptosis.[33] Blockade of these receptors is thought to provide neuroprotection in the setting of SAH.[33] Brilliant blue G (BBG) is a drug currently undergoing study as a P2X7R receptor antagonist. BBG administered after SAH resulted in improved neurologic function and reduced brain edema in rat studies.[33] Specifically in relation to neurogenic pulmonary edema, BBG was found in a rat model to reduce pulmonary edema (as determined by pulmonary index measurements) by decreasing breakdown of the cellular tight junctions in the alveolar-capillary membrane.[34] However, human studies are lacking.

Stress Cardiomyopathy

Cardiac complications following SAH are numerous and include electrocardiogram changes (abnormalities in ST and T waves, QT prolongation, U waves), ventricular and supraventricular arrhythmias, troponin level increase, and myocardial dysfunction in the absence of coronary vasospasm.[35] The incidence of arrhythmias is 35% following SAH, with roughly 5% being life threatening.[31] The reported incidence of left ventricular wall motion abnormalities is 22%, with troponin level increase found in 68% of patients with SAH.[36]

Takotsubo cardiomyopathy is a unique form of cardiac dysfunction found in patients with SAH (**Box 1**). This syndrome consists of transient wall motion abnormalities of the left ventricle that are often triggered by extreme emotional or physical stress.[37] Specifically, the apex of the left ventricle balloons outward with relative sparing of the basal segments.[38] The cause of this ventricular abnormality is unknown but proposed

---

Box 1
Mayo Clinic criteria for diagnosing apical ballooning syndrome (takotsubo or stress cardiomyopathy)

- Transient hypokinesis, akinesis, or dyskinesis of the left ventricular midsegments with or without apical involvement; the regional wall motion abnormalities extend beyond a single epicardial vascular distribution; a stressful trigger is often, but not always present[a]

- Absence of obstructive coronary disease or angiographic evidence of acute plaque rupture[b]

- New electrocardiographic abnormalities (ST-segment elevation and/or T-wave inversion) or modest increase in cardiac troponin level

- Absence of:
  o Pheochromocytoma
  o Myocarditis

[a] There are rare exceptions to these criteria, such as those patients in whom the regional wall motion abnormality is limited to a single coronary territory.
[b] It is possible that patients with obstructive coronary atherosclerosis may also develop apical ballooning syndrome. However, this is very rare in our experience and in the published literature, perhaps because such cases are misdiagnosed as an acute coronary syndrome.
From Prasad A, Lerman A, Rihal CS. Apical ballooning syndrome (Tako-Tsubo or stress cardiomyopathy): a mimic of acute myocardial infarction. Am Heart J 2008;155(3):408–17; with permission.

mechanisms include coronary microvascular dysfunction, coronary artery spasm, catecholamine-induced myocardial stunning, reperfusion injury following acute coronary syndrome, myocardial microinfarction, and abnormal cardiac fatty acid metabolism.[39] It is also postulated that catecholamine surge could be responsible for the changes in the myocardium, as shown by a greater concentration of adrenoreceptors in the left ventricular apex compared with the base.[39] Histologic evaluation of affected heart tissue reveals myocardial contraction band necrosis.[36] Cardiac MRI may be used to help assist with diagnosis because it can assess wall motion abnormalities and left ventricular ejection fraction. Cardiac MRI may also be used to differentiate takotsubo cardiomyopathy from other potential diagnoses, such as myocarditis and myocardial infarction, which show delayed gadolinium enhancement.[40]

Complications associated with takotsubo cardiomyopathy include hypotension, arrhythmias, heart failure, ventricular rupture, and thrombosis formation in the dilated left ventricular apex.[40] Treatment of takotsubo cardiomyopathy is largely supportive. Attempts should be made to minimize myocardial oxygen demand while maximizing myocardial oxygen supply. Arrhythmias should be treated promptly. Intra-aortic balloon pump counterpulsation may be needed to support severely decompensated patients.[38] Left ventricular outflow obstruction in this setting should be treated with resuscitation, beta-blockade, and blood pressure support with an alpha-agonist such as phenylephrine.[40] Although complications of takotsubo may increase morbidity, overall prognosis is good because myocardial dysfunction is typically reversible after 3 to 5 days.[38] Most patients show improved systolic function in as little as 1 week, with complete recovery in 3 to 4 weeks.[40]

Pituitary Dysfunction

Pituitary dysfunction in the setting of SAH is underappreciated and can continue to affect patients' quality of life long after the acute phase of recovery. Patients recovering from SAH often report symptoms of fatigue and impaired memory and planning, which are symptoms of growth hormone deficiency. Khajeh and colleagues[41] prospectively evaluated pituitary dysfunction in SAH at a single institution. In their study of 88 patients, 39% of survivors of SAH showed at least 1 pituitary hormone deficiency, with gonadotropin and/or growth hormone deficiency occurring more frequently than adrenocorticotrophic hormone or thyroid-stimulating hormone. Hydrocephalus was identified as an independent predictor for persistent pituitary dysfunction at 6 months. As such, restoration of normal anterior pituitary hormone levels and/or timely treatment of hydrocephalus could potentially improve rehabilitation and may warrant screening in select patients.[42]

Disorders of Sodium

Sodium dysregulation is common in patients with SAH, with hyponatremia occurring most commonly followed by hypernatremia and then mixed presentation. Hyponatremia may be caused by cerebral salt wasting (CSW) or syndrome of inappropriate antidiuretic hormone (SIADH). Although fluid restriction is typical in the management of SIADH, it is inappropriate in the setting of SAH, because hypovolemia and negative fluid balance worsen the outcome.[35]

Hypernatremia is associated with increased mortality, poor outcome, and acute kidney injury.[43–45] Usually secondary to osmotic therapy for intracranial hypertension, such as hypertonic saline and mannitol, hypernatremia may also be caused by diabetes insipidus (DI). In addition, fludrocortisone may be useful to maintain a positive fluid and sodium balance (**Box 2**).

Box 2
Sodium dysfunction in SAH: diagnosis and treatment summary

DI

- ↑ UOP (>250 mL/h)
- Dilute urine (urine SG<1.005)
- N or ↑ serum Na
- Vasopressin or desamino-D-arginine vasopressin
- Vigorous hydration with half-normal saline to euvolemia

SIADH

- Urine Na greater than 18
- ↓ Na + Osm
- Na less than 134, Osm less than 280
- HTS + loop diuretics
- Demeclocycline
- Conivaptan and tolvaptan

CSW

- Hypovolemia
- ↓ Na
- ↑ K^+ supports CSW more than SIADH
- Avoid fluid restriction
- ↑ Plasma BUN/Cr

Abbreviations: ↑, increased; ↓, decreased; BUN, blood urea nitrogen; Cr, creatinine; HTS, hypertonic saline; N, normal; Osm, serum osmolality; SG, specific gravity; UOP, urine output.

DELAYED CEREBRAL ISCHEMIA

DCI is defined as the occurrence of focal neurologic impairment or a decrease of at least 2 points on the Glasgow Coma Scale lasting at least 1 hour and not attributable to other causes. DCI occurs in up to 40% of patients surviving the initial hemorrhage and is the primary cause of mortality if it leads to infarction. Historically, DCI was thought to be caused by cerebral vasospasm, and the terms were used nearly interchangeably. However, although cerebral vasospasm can contribute to DCI, the two phenomena are not equivalent. The term vasospasm should be reserved for those patients with angiographic evidence of arterial narrowing and this condition can be screened for with noninvasive daily transcranial Doppler assessment of cerebral blood flow velocities.[46] Although vasospasm is one cause of DCI, other pathophysiologic mechanisms occur. In 70% of patients with SAH, angiographic evidence of vasospasm is seen, but only up to 40% develop DCI. Furthermore, regions of hypoperfusion are observed in patients without concomitant angiographic vasospasm.[47]

Cortical Spreading Depolarization

Cortical spreading depolarization (CSD) is described as self-propagating tissue depolarization waves that lead to ischemia and are associated with increase in extracellular potassium levels, compromise of the blood-brain barrier, formation of vasogenic and cytotoxic edema, and change in regional cerebral blood flow.[46,48] During these waves of depolarization, a failure of homeostasis leads to neuronal swelling and increased metabolic rate, resulting in acidosis. Regional cerebral blood flow changes then occur in 2 phases: first, a hyperemic response with vasodilation up to 2 minutes; then a longer, reactive vasoconstrictive epoch lasting up to 2 hours and leading to oligemia. Described as an inverse neurovascular response, vasoconstriction persists even during a depolarization wave with associated increased metabolic needs. Although the processes involved in CSD are not entirely understood, some overlap with cerebral vasospasm does occur. The time course is similar, occurring on approximately post-bleed day 4 to 10.[49] However, Woitzik and colleagues[50] showed that nicardipine pellets reduced angiographic vasospasm, but not CSD. Ketamine has been associated with reduced CSD in patients with brain injury and warrants further investigation.

Microthrombi

On aneurysm rupture, an inflammatory reaction is initiated involving reduced nitric oxide levels and increased fibrin and platelet aggregation leading to a hypercoagulable state. Microthrombi may form and contribute to DCI. In this setting, the antifibrinolytic, tranexamic acid, as described earlier, was associated with increased ischemic complications when used for blood clot stabilization before aneurysm obliteration. More broadly, experimentation with anticoagulants may yet hold promise in improving outcomes in SAH. One study of enoxaparin administration showed more intracranial hemorrhages but no change in 3-month outcome.[51] However, another study using low-dose enoxaparin showed improved outcome and less DCI.[52] Further studies are needed.

Dysfunctional Cerebral Autoregulation

Defective cerebral autoregulation is seen early after SAH. Measures of dynamic cerebral autoregulation (see Kirkman MA, Smith M: Multimodality neuromonitoring, in this issue) have been correlated with the development of angiographic vasospasm and, more critically, to the development of DCI. Impaired cerebral autoregulation can be detected before changes in mean blood flow velocities by transcranial Doppler. This early detection can be used to optimize individual therapy for DCI. With continued clinical investigation, scores may be developed that can be used to assess risk of vasospasm and DCI and help define optimal cerebral perfusion pressures.[53–55]

THERAPIES
Euvolemic Hypertensive Therapy

Hyperdynamic, hypertensive, hemodilutional therapy (triple H) is no longer consistent with the current understanding of SAH management and DCI. Insufficient data exist to recommend iatrogenic hemodilution, and prophylactic hypervolemia via fluid administration may worsen cardiopulmonary status. However, euvolemia maintenance is critical, and it seems that hypertension may be the most important H. Therefore, euvolemic hypertensive therapy is currently recommended for patients with DCI.[56] Although clinical experience and anecdotes report reversal of signs of DCI with initiation of hypertension, randomized trials are still needed. The Hypertension Induction in the Management of Aneurysmal Subarachnoid Haemorrhage with Secondary Ischaemia (HIMALAIA) trial is a multicenter randomized controlled trial that will compare hypertension induction with no blood pressure augmentation in the treatment of DCI. The primary focus will be on outcome, but CT perfusion studies assessing changes in cerebral blood flow will also be performed and will likely refine the understanding of DCI.[57]

Calcium Channel Blockers

Oral nimodipine therapy is the mainstay of SAH treatment. Based largely on the landmark trial published in 1983 evaluating oral nimodipine,[58] patients with SAH should be given 60 mg orally every 4 hours for 21 days to reduce the rate of DCI and improve outcome. Nimodipine administration remains the only class 1, level A recommendation in the management of SAH. In addition to its vasodilating effects mediated through inhibition of L-type calcium channels, nimodipine also seems to offer neuroprotection through fibrinolytic activity and inhibition of CSD. Another dihydropyridine calcium channel blocker, nicardipine, is used extensively in patients with SAH for intravenous blood pressure control. Haley and colleagues[59] showed reduced angiographic vasospasm, but no improvement in outcomes with intravenous nicardipine. Administration of intraventricular and intra-arterial nicardipine has also shown reduced evidence of vasospasm.[55]

Statins

SAH and DCI involve a complex and incompletely understood inflammatory cascade. 3-Hydroxy-3-methylglutaryl coenzyme A reductase inhibitors (statins) have many effects that may be beneficial in SAH:

- Improve endothelial vasomotor function
- Increase nitric oxide availability
- Act as antioxidants
- Induce angiogenesis
- Suppress cytokines in cerebral ischemia

Statin therapy in SAH was shown to decrease DCI in small trials of pravastatin. However, recent larger trials failed to show benefit. The Simvastatin in Aneurysmal Subarachnoid Hemorrhage (STASH)[60] trial evaluated 800 patients given 40 mg of simvastatin versus placebo for 21 days and was unable to show improved short-term or long-term outcomes. Pilot trials used higher-dose simvastatin, so 80 mg of simvastatin were compared with 40 mg but still failed to show any benefit in the High-dose Simvastatin for Aneurysmal SAH (HDS-SAH) trial.[61] Therefore, initiation of statin therapy after SAH cannot be recommended for routine administration in SAH care. The Neurocritical Care Society recommends that patients taking statins before SAH have their medication continued, although this recommendation is based on low-quality evidence.[61] The investigators' practice is to continue but not initiate statin therapy in SAH.

Glucose Management

Dysglycemia following SAH has been associated with poor clinical outcomes,[62] including DCI, and was reported in the literature as early as 1925.[63] The most recently published guidelines address hyperglycemia as a risk factor for complications.[62–64] Association between hyperglycemia and poor SAH outcome may relate to the progression from ischemia to infarction, caused by increased inflammatory response, increased vasoconstriction, and increased coagulation with decreased fibrinolysis.[65]

Several mechanisms contribute to the occurrence of hyperglycemia in SAH: activation of the sympathetic autonomic nervous system, resulting in increased levels of stress hormones (eg, cortisol, catecholamines) and hormones of the hypothalamic-pituitary-adrenal axis; release of cytokines, accompanying an increased inflammatory response; and hypothalamic dysfunction.[66]

Cerebral microdialysis can be an effective neuromonitoring technique for the detection of metabolic disruption.[65] More data are needed to more precisely define the role of glucose management in neuroprotection.[67,68]

SUMMARY

SAH remains a cerebrovascular disease with high morbidity and mortality. Outcomes can be improved when care is provided at high-volume centers with dedicated neuro-intensive care units and specialized staff. Trials relating to virtually every component of management are still ongoing, with an emphasis on neuroprotective mechanisms and therapies. Through deliberate practice, adherence to evolving best evidence, and developing understanding of secondary brain injury, progress will be made.

ACKNOWLEDGMENTS

The authors acknowledge the assistance of Christopher Hayes for his graphic design work, and Emily Topmiller for her assistance in article preparation.

REFERENCES

1. de Rooij NK, Linn FH, van der Plas JA, et al. Incidence of SAH; a systemic review with emphasis on region, age, gender, and time trends. J Neurol Neurosurg Psychiatry 2007;78:1365–72.
2. Qureshi AI, Palesch YY, Martin R, et al. Interpretation and Implementation of Intensive Blood Pressure Reduction in Acute Cerebral Hemorrhage Trial (INTERACT II). J Vasc Interv Neurol 2014;7(2):34–40.
3. Butcher K, Jeerakathil T, Emery D, et al. The Intracerebral Haemorrhage Acutely Decreasing Arterial Pressure Trial: ICH ADAPT. Int J Stroke 2010;5(3):227–33.
4. Woloszyn AV, McAllen KJ, Figueroa BE, et al. Retrospective evaluation of nicardipine versus labetalol for blood pressure control in aneurysmal subarachnoid hemorrhage. Neurocrit Care 2012;16(3):376–80.
5. Qureshi AI, Palesch YY. Antihypertensive Treatment of Acute Cerebral Hemorrhage (ATACH) II: design, methods, and rationale. Neurocrit Care 2011;15(3):559–76.
6. Kassell NF, Torner JC. Aneurysmal rebleeding: a preliminary report from the Cooperative Aneurysm Study. Neurosurgery 1983;13:479–81.
7. de Gans K, Nieuwkamp DJ, Rinkel GJ, et al. Timing of aneurysm surgery in subarachnoid hemorrhage: a systematic review of the literature. Neurosurgery 2002;50(2):336–40.
8. Phillips TJ, Dowling RJ, Yan B, et al. Does treatment of ruptured intracranial aneurysms within 24 hours improve clinical outcome? Stroke 2011;42(7):1936–45.
9. Siddiq F, Chaudhry SA, Tummala RP, et al. Factors and outcomes associated with early and delayed aneurysm treatment in subarachnoid hemorrhage patients in the United States. Neurosurgery 2012;71(3):670–7.
10. Schuette AJ, Barrow DL. Treatment of ruptured middle cerebral aneurysms associated with intracerebral hematomas. World Neurosurg 2013;80(3–4):266–7.
11. Gaberel T, Magheru C, Emery E, et al. Antifibrinolytic therapy in the management of aneurismal subarachnoid hemorrhage revisited. A meta-analysis. Acta Neurochir (Wien) 2012;154(1):1–9.
12. Germans MR, Post R, Coert BA, et al. Ultra-early tranexamic acid after subarachnoid hemorrhage (ULTRA): study protocol for a randomized controlled trial. Trials 2013;14:143.

13. Molyneux AJ, Birks J, Clarke A, et al. The durability of endovascular coiling versus neurosurgical clipping of ruptured cerebral aneurysms: 18 year follow-up of the UK cohort of the International Subarachnoid Aneurysm Trial (ISAT). Lancet 2015;385(9969):691-7.

14. Spetzler RF, McDougall CG, Zabramski JM, et al. The barrow ruptured aneurysm trial: 6-year results. J Neurosurg 2015;123(3):609-17.

15. Chalouhi N, Whiting A, Anderson EC, et al. Comparison of techniques for ventri-culoperitoneal shunting in 523 patients with subarachnoid hemorrhage. J Neurosurg 2014;121(4):904-7.

16. Molyneux AJ, Kerr RS, Yu LM, et al, International Subarachnoid Aneurysm Trial (ISAT) Collaborative Group. International Subarachnoid Aneurysm Trial (ISAT) of neurosurgical clipping versus endovascular coiling in 2143 patients with ruptured intracranial aneurysms: a randomised comparison of effects on survival, dependency, seizures, rebleeding, subgroups, and aneurysm occlusion. Lancet 2005;366(9488):809-17.

17. Guglielmi G, Viñuela F, Dion J, et al. Electrothrombosis of saccular aneurysms via endovascular approach. Part 2: Preliminary clinical experience. J Neurosurg 1991;75(1):8-14.

18. Standhardt H, Boecher-Schwarz H, Gruber A, et al. Endovascular treatment of unruptured intracranial aneurysms with Guglielmi detachable coils: short- and long-term results of a single-centre series. Stroke 2008;39(3):899-904.

19. Moret J, Cognard C, Weill A, et al. The "remodelling technique" in the treatment of wide neck intracranial aneurysms. angiographic results and clinical follow-up in 56 cases. Interv Neuroradiol 1997;3(1):21-35.

20. Pierot L, Spelle L, Vitry F, ATENA Investigators. Immediate clinical outcome of patients harboring unruptured intracranial aneurysms treated by endovascular approach: results of the ATENA study. Stroke 2008;39(9):2497-504.

21. Pierot L, Cognard C, Spelle L, et al. Safety and efficacy of balloon remodeling technique during endovascular treatment of intracranial aneurysms: critical review of the literature. AJNR Am J Neuroradiol 2012;33(1):12-5.

22. Peluso JP, van Rooij WJ, Sluzewski M, et al. A new self-expandable nitinol stent for the treatment of wide-neck aneurysms: initial clinical experience. AJNR Am J Neuroradiol 2008;29(7):1405-8.

23. Piotin M, Blanc R, Spelle L, et al. Stent-assisted coiling of intracranial aneurysms: clinical and angiographic results in 216 consecutive aneurysms. Stroke 2010; 41(1):110-5.

24. Biondi A, Janardhan V, Katz JM, et al. Neuroform stent-assisted coil embolization of wide-neck intracranial aneurysms: strategies in stent deployment and midterm follow-up. Neurosurgery 2007;61(3):460-8.

25. Sedat J, Chau Y, Mondot L, et al. Endovascular occlusion of intracranial wide-necked aneurysms with stenting (Neuroform) and coiling: mid-term and long-term results. Neuroradiology 2009;51(6):401-9.

26. Lubicz B, Bandeira A, Bruneau M, et al. Stenting is improving and stabilizing anatomical results of coiled intracranial aneurysms. Neuroradiology 2009;51(6):419-25.

27. Place holder for pipeline embolization device approval {&*&*(&(}. Available at: www.fda.gov. Accessed April 06, 2011.

28. Nelson PK, Lylyk P, Szikora I, et al. The pipeline embolization device for the intracranial treatment of aneurysms trial. AJNR Am J Neuroradiol 2011;32(1):34-40.

29. Lylyk P, Miranda C, Ceratto R, et al. Curative endovascular reconstruction of cerebral aneurysms with the pipeline embolization device: the Buenos Aires experience. Neurosurgery 2009;64(4):632-42.

30. Busl KM, Bleck TP. Neurogenic pulmonary edema. Crit Care Med 2015;43(8): 1710–5.
31. Bruder N, Rabinstein A. Participants in the international multi-disciplinary consensus conference on the critical care management of subarachnoid hemorrhage. Cardiovascular and pulmonary complications of aneurysmal subarachnoid hemorrhage. Neurocrit Care 2011;15(2):257–69.
32. Baker M, Bastin MT, Cook AM, et al. Hypoxemia associated with nimodipine in a patient with an aneurysmal subarachnoid hemorrhage. Am J Health Syst Pharm 2015;72(1):39–43.
33. Chen S, Ma Q, Krafft PR, et al. P2X7 receptor antagonism inhibits p38 mitogen-activated protein kinase activation and ameliorates neuronal apoptosis after subarachnoid hemorrhage in rats. Crit Care Med 2013;41(12):e466–74.
34. Chen S, Zhu Z, Klebe D, et al. Role of P2X purinoceptor 7 in neurogenic pulmonary edema after subarachnoid hemorrhage in rats. PLoS One 2014;9(2):e89042.
35. D'Souza S. Aneurysmal subarachnoid hemorrhage. J Neurosurg Anesthesiol 2015;27(3):222–40.
36. Wartenberg KE, Mayer SA. Medical complications after subarachnoid hemorrhage. Neurosurg Clin North Am 2010;21(2):325–38.
37. Kurisu S, Kihara Y. Tako-tsubo cardiomyopathy: clinical presentation and underlying mechanism. J Cardiol 2012;60(6):429–37.
38. Lee K, Choi HA, Edwards N, et al. Perioperative critical care management for patients with aneurysmal subarachnoid hemorrhage. Korean J Anesthesiol 2014; 67(2):77–84.
39. Davison DL, Chawla LS, Selassie L, et al. Neurogenic pulmonary edema: successful treatment with IV phentolamine. Chest 2012;141(3):793–5.
40. Komamura K, Fukui M, Iwasaku T, et al. Takotsubo cardiomyopathy: Pathophysiology, diagnosis and treatment. World J Cardiol 2014;6(7):602–9.
41. Khajeh L, Blijdorp K, Heijenbrok-Kal MH, et al. Pituitary dysfunction after aneurysmal subarachnoid haemorrhage: course and clinical predictors—the HIPS study. J Neurol Neurosurg Psychiatry 2015;86(8):905–10.
42. Jovanovic V, Pekic S, Stojanovic M, et al. Neuroendocrine dysfunction in patients recovering from subarachnoid hemorrhage. Hormones (Athens) 2010;9(3): 235–44.
43. Lantigua H, Ortega-Gutierrez S, Schmidt JM, et al. Subarachnoid hemorrhage: who dies, and why? Crit Care 2015;19:309.
44. Kumar AB, Shi Y, Shotwell MS, et al. Hypernatremia is a significant risk factor for acute kidney injury after subarachnoid hemorrhage: a retrospective analysis. Neurocrit Care 2015;22(2):184–91.
45. Beseoglu K, Etminan N, Steiger HJ, et al. The relation of early hypernatremia with clinical outcome in patients suffering from aneurysmal subarachnoid hemorrhage. Clin Neurol Neurosurg 2014;123:164–8.
46. Sánchez-Porras R, Zheng Z, Santos E, et al. The role of spreading depolarization in subarachnoid hemorrhage. Eur J Neurol 2013;20(8):1121–7.
47. Vergouwen MD, Vermeulen M, van Gijn J, et al. Definition of delayed cerebral ischemia after aneurysmal subarachnoid hemorrhage as an outcome event in clinical trials and observational studies: proposal of a multidisciplinary research group. Stroke 2010;41(10):2391–5.
48. Dreier JP. The role of spreading depression, spreading depolarization and spreading ischemia in neurological disease. Nat Med 2011;17(4):439–47.
49. Hertle DN, Dreier JP, Woitzik J, et al, Cooperative Study of Brain Injury Depolarizations (COSBID). Effect of analgesics and sedatives on the occurrence of

spreading depolarizations accompanying acute brain injury. Brain 2012;135(Pt 8):2390–8.

50. Woitzik J, Dreier JP, Hecht N, et al, COSBID Study group. Delayed cerebral ischemia and spreading depolarization in absence of angiographic vasospasm after subarachnoid hemorrhage. J Cereb Blood Flow Metab 2012;32(2):203–12.

51. Siironen J, Juvela S, Varis J, et al. No effect of enoxaparin on outcome of aneurysmal subarachnoid hemorrhage: a randomized, double-blind, placebo-controlled clinical trial. J Neurosurg 2003;99(6):953–9.

52. Wurm G, Tomancok B, Nussbaumer K, et al. Reduction of ischemic sequelae following spontaneous subarachnoid hemorrhage: a double-blind, randomized comparison of enoxaparin versus placebo. Clin Neurol Neurosurg 2004;106(2): 97–103.

53. Bijlenga P, Czosnyka M, Budohoski KP, et al. Optimal cerebral perfusion pressure" in poor grade patients after subarachnoid hemorrhage. Neurocrit Care 2010;13(1):17–23.

54. Budohoski KP, Czosnyka M, Smielewski P, et al. Impairment of cerebral autoregulation predicts delayed cerebral ischemia after subarachnoid hemorrhage: a prospective observational study. Stroke 2012;43(12):3230–7.

55. Otite F, Mink S, Tan CO, et al. Impaired cerebral autoregulation is associated with vasospasm and delayed cerebral ischemia in subarachnoid hemorrhage. Stroke 2014;45(3):677–82.

56. Dankbaar JW, Slooter AJ, Rinkel GJ, et al. Effect of different components of triple-H therapy on cerebral perfusion in patients with aneurysmal subarachnoid haemorrhage: a systematic review. Crit Care 2010;14(1):R23.

57. Gathier CS, van den Bergh WM, Slooter AJ, HIMALAIA-Study Group. HIMALAIA (Hypertension Induction in the Management of AneurysmaL subArachnoid haemorrhage with secondary IschaemiA): a randomized single-blind controlled trial of induced hypertension vs. no induced hypertension in the treatment of delayed cerebral ischemia after subarachnoid hemorrhage. Int J Stroke 2014;9(3): 375–80.

58. Allen GS, Ahn HS, Preziosi TJ, et al. Cerebral arterial spasm-a controlled trial of nimodipine in patients with subarachnoid hemorrhage. N Engl J Med 1983;308: 619–24.

59. Haley EC Jr, Kassell NF, Torner JC. A randomized trial of nicardipine in subarachnoid hemorrhage: angiographic and transcranial Doppler ultrasound results. A report of the Cooperative Aneurysm Study. J Neurosurg 1993;78(4):548–53.

60. Turner CL, Budohoski K, Smith C, et al, STASH Collaborators. Elevated baseline C-reactive protein as a predictor of outcome after aneurysmal subarachnoid hemorrhage: data from the Simvastatin in Aneurysmal Subarachnoid Hemorrhage (STASH) trial. Neurosurgery 2015;77(5):786–93.

61. Diringer MN, Bleck TP, Claude Hemphill J 3rd, et al, Neurocritical Care Society. Critical care management of patients following aneurysmal subarachnoid hemorrhage: recommendations from the Neurocritical Care Society's Multidisciplinary Consensus Conference. Neurocrit Care 2011;15(2):211–40.

62. Kruyt ND, Biessels GJ, de Haan RJ, et al. Hyperglycemia and clinical outcome in aneurysmal subarachnoid hemorrhage: a meta-analysis. Stroke 2009;40(6): e424–30.

63. Bederson JB, Connolly ES Jr, Batjer HH, et al, American Heart Association. Guidelines for the management of aneurysmal subarachnoid hemorrhage: a statement for healthcare professionals from a special writing group of the Stroke Council, American Heart Association. Stroke 2009;40(3):994–1025.

64. Connolly ES Jr, Rabinstein AA, Carhuapoma JR, et al, American Heart Association Stroke Council, Council on Cardiovascular Radiology and Intervention, Council on Cardiovascular Nursing, Council on Cardiovascular Surgery and Anesthesia, Council on Clinical Cardiology. Guidelines for the management of aneurysmal subarachnoid hemorrhage: a guideline for healthcare professionals from the American Heart Association/American Stroke Association. Stroke 2012;43(6):1711–37.

65. Rostami E. Glucose and the injured brain-monitored in the neurointensive care unit. Front Neurol 2014;5:91.

66. Kruyt ND, Biessels GJ, DeVries JH, et al. Hyperglycemia in aneurysmal subarachnoid hemorrhage: a potentially modifiable risk factor for poor outcome. J Cereb Blood Flow Metab 2010;30(9):1577–87.

67. Badenes R, Gruenbaum SE, Bilotta F. Cerebral protection during neurosurgery and stroke. Curr Opin Anaesthesiol 2015;28(5):532–6.

68. Sonneville R, Vanhorebeek I, den Hertog HM, et al. Critical illness-induced dysglycemia and the brain. Intensive Care Med 2015;41(2):192–202.

Neuromuscular Disease in the Neurointensive Care Unit

Veronica Crespo, MD[a], Michael L. "Luke" James, MD[a,b],*

KEYWORDS

- Neuromuscular diseases • Acute inflammatory demyelinating polyneuropathy
- Guillain-Barre syndrome • Amyotrophic lateral sclerosis • Myasthenia gravis
- Critical illness polyneuropathy • Critical illness myopathy

KEY POINTS

- Knowledge about the management of critically ill patients with neuromuscular disease is required for anesthesiologists, because these patients are frequently encountered in the intensive care unit and operating room.
- Acute inflammatory demyelinating polyneuropathy, that is, Guillain-Barre syndrome, is a rapidly progressive peripheral neuropathy that may affect multiple organ systems. Mainstay therapies include intravenous immunoglobulins or plasma exchange.
- The most common cause of death from amyotrophic lateral sclerosis is respiratory failure. This disease requires special attention to airway management and prevention of aspiration.
- Myasthenia gravis encompasses multiple known subtypes, all of which respond differently to the variety of available treatments. If symptomatic treatment with an acetylcholinesterase inhibitor is unsatisfactory, immunosuppressive therapy should be started.
- Critical illness polyneuropathy and myopathy are conditions of significant morbidity, including prolonged mechanical ventilation. Management is mainly preventative and supportive.

NEUROMUSCULAR DISEASE IN THE NEUROINTENSIVE CARE UNIT

Neuromuscular diseases are different syndromes that affect nerve, muscle, and/or neuromuscular junction (**Fig. 1, Table 1**). Afflictions encompassed in this category present a myriad of challenges for anesthesiologists in the operating room, pain clinic, and intensive care unit (ICU). These challenges and advances in management will be reviewed for neuromuscular diseases most commonly encountered.

This work is published in collaboration with the Society for Neuroscience in Anesthesiology and Critical Care.
Disclosures: None.
[a] Department of Anesthesiology, Duke University, Erwin Road, Durham, NC 27710, USA;
[b] Department of Neurology, Duke University, Erwin Road, Durham, NC 27710, USA
* Corresponding author. Department of Anesthesiology, Duke University Medical Center, Box 3094 DUMC, 2031 Erwin Road, Room 5688 HAFS, Durham, NC 27710.
E-mail address: michael.james@duke.edu

Anesthesiology Clin 34 (2016) 601–619
http://dx.doi.org/10.1016/j.anclin.2016.04.010
1932-2275/16/$ – see front matter © 2016 Elsevier Inc. All rights reserved.

anesthesiology.theclinics.com

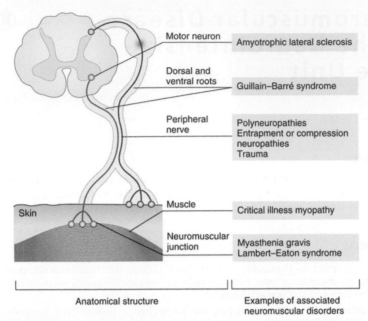

Fig. 1. Sites of neuromuscular disease in the spinal cord, peripheral nerves, and muscle. (*Adapted from* Horlings CG, van Engelen BG, Allum JH, et al. A weak balance: the contribution of muscle weakness to postural instability and falls. Nat Clin Pract Neurol 2008;4:504–15; with permission.)

Anesthetic Considerations

The anesthetic plan for any individual involves thoughtfully weighing risks against benefits, often from a range of choices. When the potential risks and benefits of possible interventions are not well known or misunderstood, decisions regarding the appropriate course of action are fraught with the possibility of unintended consequences. One such instance is the decision to provide general versus neuraxial or regional anesthesia to a patient with neuromuscular disease. General anesthesia with endotracheal intubation provides unconsciousness with a secure airway. However, general anesthesia often requires the use of neuromuscular blocking agents (NMBA); thus, removal of the endotracheal tube after the end of surgery may be challenging in patients with advanced neuromuscular disease, especially those with pre-existing respiratory or bulbar involvement. Neuraxial techniques generally preclude endotracheal tube placement but may cause profound respiratory impairment in patients with neuromuscular disease with only minimal involvement of accessory respiratory muscles. Perhaps the greatest concern regarding regional (including neuraxial) anesthesia in this patient population is the uncertainty of its effects on disease progression and symptoms. Regional anesthesia has been documented as safe and without neurologic sequelae in the setting of pre-existing neuromuscular disease by multiple sources.[1–3] However, neuromuscular disease exacerbations have been reported, although possibly influenced by a variety of confounding factors.[3–5] A causal link between regional anesthesia and neuromuscular disease exacerbation has not been established.

Regardless of anesthetic type, patients with neuromuscular disease may present to the ICU after surgery and anesthesia. Preoperative optimization of the patient may greatly decrease the need for postoperative mechanical ventilation and improve

Table 1 Neuromuscular diseases			
	Pathophysiology	**Diagnosis**	**Treatment**
AIDP	Immunologic	• Clinical features • Raised protein concentration in CSF • Neurophysiological studies[7]	• IVIg or PLEX • Supportive
ALS	Unknown, but mutations in c9orf72, TDP-43, FUS, SOD-1; glutamate-mediated excitotoxicity have been implicated	• Clinical features • Electrophysiological confirmation[26]	• Riluzole • Noninvasive and invasive ventilation support as needed • Other supportive therapies
MG	Autoimmune, antibodies target the postsynaptic membrane of the NMJ	• Antibodies against AChR, MUSK, or LRP4 • Neurophysiologic testing if antibodies are undetectable	• Pyridostigmine • Steroids • Azathioprine • Other immuno-suppressive agents • Thymectomy in thymoma • IVIg or PLEX for exacerbations
Critical illness polyneuropathy/CIM	Likely the same metabolic, cellular, and microcirculatory mechanisms as multiorgan dysfunction[5]	• Clinical features • Neurophysiologic testing • Muscle biopsy if needed to distinguish CIM from other myopathies	• Preventative (intensive insulin therapy) • Supportive • Treat and prevent sepsis

Abbreviations: CSF, cerebrospinal fluid; NMJ, neuromuscular junction.

pain management. During general anesthesia, conservative dosing of intravenous and volatile anesthetics as well as opioids and sedatives may also improve postoperative respiratory function. Patients with neuromuscular disease may be best suited to the use of rapidly reversible opioids and anesthetics.[6] Patients with denervation or prolonged paralysis may develop extrajunctional acetylcholine receptors (AChR). For this reason, succinylcholine may trigger fatal hyperkalemia and should be avoided. Meanwhile, the same mechanism causes increased resistance to nondepolarizing NMBAs; thus, larger doses of these agents may be warranted.[1] Of course, ultimately, the need for paralysis should be cautiously weighed against the potential for prolonged weakness. Adequate reversal of NMBA should be confirmed with train-of-4 monitoring, and removal of the endotracheal tube should take place once the patient is completely awake.[6] Myasthenia gravis (MG) is notably different regarding NMBAs, given neuromuscular junction involvement. Additional anesthetic considerations specific to this disease are reviewed in **Box 1**.

Ultrasound guidance while performing regional blocks is advocated to prevent complications and allow for smaller doses of local anesthetics.[2] During regional anesthesia, patients should be closely monitored for signs and symptoms of impaired ventilation and/or oxygenation. The increased risk of aspiration on some patients with neuromuscular disease should be taken into consideration when administering sedation.

> **Box 1**
> **Additional anesthetic considerations in myasthenia gravis**
>
> *Preoperative management controversy*
>
> - Continuing AChE-I preoperatively:
>   - Improves muscle strength
>   - Improves patient comfort
>   - Decreases risk of respiratory stress
>
> - Discontinuing AChE-I preoperatively:
>   - Avoids reduced cholinesterase activity
>   - Facilitates quicker onset and smaller doses of nondepolarizing NMBA
>   - May decrease the risk of vagal responses
>
> - Therapy, such as a cycle of PLEX, should be attempted preoperatively in patients with acute exacerbations who must undergo surgery urgently
>
> - If surgery during an acute exacerbation can be postponed, it should be, because surgery is best performed when patients are in their optimal state
>
> *Effects of neuromuscular blocking agents*
>
> - Patients who are not taking pyridostigmine:
>   - Are resistant to succinylcholine, due to reduced number of AChR
>   - Are sensitive to nondepolarizing NMBAs, due to reduced number of AChR
>
>   - Patients who are taking pyridostigmine:
>   - May have prolonged succinylcholine and ester local anesthetics effect, due to inhibition of plasma cholinesterase (PLEX may also have this effect)
>   - May experience slower onset and require higher doses of nondepolarizing NMBAs. Reversal may be difficult if AChE has already been maximally inhibited
>
> Increased sensitivity to the muscle relaxation of volatile anesthetics, for which train of 4 monitoring is recommended.
>
> Suboptimal pain control may precipitate a myasthenic crisis. Opioids should be short-acting and titrated carefully, given its effects on respiratory and gastrointestinal function. Regional anesthesia is an alternative for postoperative pain management.
>
> *Data from* Romero A, Joshi GP. Neuromuscular disease and anesthesia. Muscle Nerve 2013;48(3):451–60; and Blichfeldt-Lauridsen L, Hansen BD. Anesthesia and myasthenia gravis. Acta Anaesthesiol Scand 2012;56(1):17–22.

ACUTE INFLAMMATORY DEMYELINATING POLYNEUROPATHY
Overview

Acute inflammatory demyelinating polyneuropathy (AIDP), commonly known as Guillain-Barre syndrome, affects peripheral nerve axons via demyelination or direct axonal injury. Characterized by an acute onset of rapidly progressive and symmetric, ascending peripheral weakness, AIDP most commonly presents with lower extremity weakness in the setting of hyporeflexia or areflexia. Commonly preceded by an infectious process, underlying immunologic pathophysiology is suspected.[7] In distinction to chronic inflammatory demyelinating polyneuropathy (CIDP), the acute phase of AIDP may consist of respiratory failure and autonomic dysfunction, and neuropathic pain may ensue over the first 4 weeks after symptom onset.

Immunotherapy

Current mainstay therapies include plasma exchange (PLEX) or intravenous immunoglobulin (IVIg) to reduce recovery time. **Table 2** summarizes differences between and uses of IVIg and PLEX. The American Academy of Neurology recommends that PLEX

Table 2
Intravenous immunoglobulin and plasma exchange summary

	IVIg	PLEX
What is it?	Pooled polyclonal IgG derived from thousands of donors' serum	Procedure that centrifuges or filters blood to separate and remove plasma from cellular components
Mechanism of action	Potential anti-inflammatory and immunomodulatory pathways include • Blockade of cellular receptors • Cytokines, complement and auto-antibodies neutralization • Blocking autoantibody from binding to FcγR or modulation of immune effector cells and B cells via upregulation/downregulation of FcγRI/FcγRIIB • Saturation of FcRn	Potential therapeutic mechanisms include • Removal of plasma containing pathologic substances • Sensitization of lymphocytes to immunosuppressant and chemotherapeutic agents via alterations in their proliferation and function • Alterations in B and T cells
What neuropathies does it treat?	• FDA-approved for CIDP, MMN • "Off-label" uses with evidence from controlled trials: AIDP, MG, dermatomyositis, stiff-person syndrome, and others	AIDP, CIDP, paraproteinemic polyneuropathies, MG, and others
Therapy administration	Typically 2 g/kg IV divided in 2–5 daily doses. Maximum dose is arbitrary. If responsive to IVIg, maintenance dose (typically 1 g/kg/mo) is required, but dose and frequency should be determined by the efficacy of treatment and the duration of its benefits	Typically 1–1.5 plasma volumes exchanged per procedure per day, 3–5 d. May be prolonged for chronic disease, depending on the patient's response. Removed filtrate is replaced with colloid or combination colloid/crystalloid solutions
Adverse effects	• Headache, fever, hypertension, chills, nausea, myalgia • Chest or back pain at initiation of infusion (management includes discontinuing and restarting 30 min later at a lower rate) • Fatigue, fever, nausea after infusion up to 24 h • Thromboembolic events • Severe headache secondary to aseptic meningitis (~24–48 h) • Migraine in patients with a history • Skin reactions up to 30 d, may occur 2–5 d postinfusion, more common with high doses • Allergic reactions are not more common in IgA deficiency by itself • ATN is rare, but creatinine may increase 1–10 d after infusion and back to baseline in 2–60 d	• Paresthesias secondary to hypocalcemia, caused by citrate administered to prevent coagulation in the apheresis device • Urticaria • Pruritus • Hypotension • Tachycardia • Fever • Nausea • Vomiting

(continued on next page)

	IVIg	PLEX
Table 2 **(*continued*)**		
	IVIg	**PLEX**
Advantages/ disadvantages	• Considered safe • More convenient, simple, and comfortable for the patient • May be infused via peripheral cannula • Expensive (data regarding cost-effectiveness as compared with PLEX is limited)	• Considered safe • May not be more readily available in facilities with poor resources • May require central line placement or 2 large-bore catheters that withstand the required pressures and flows for multiple cycles • Expensive (data regarding cost-effectiveness as compared with IVIg is limited)

Abbreviations: AIDP, Guillain-Barre syndrome; ATN, acute tubular necrosis; FcγR, Fcγ receptor; FcγRI, Fcγ receptor I; FcγRIIB, Fcγ receptor IIB; FcRn, neonatal Fc receptor; FDA, US Food and Drug Administration; IgG, immunoglobulin G; IV, intravenous; MMN, multifocal motor neuropathy.

Data from Refs.[81–86]

be started within 4 weeks of onset of neuropathy in nonambulating patients and within 2 weeks of onset of neuropathy in ambulating patients.[2] This recommendation is supported by meta-analysis[8] of 5 trials comparing PLEX to standard supportive care.[9–13] These trials all reported more patients with one or more disability grade improvement at 4 weeks in the PLEX-treatment group than in the control group (57.1 vs 34.9%, respectively; relative risk = 1.64, 95% confidence interval [CI] 1.37–1.96 $P<.00001$).[8] The mean grade improvement for 4 of these trials is 1.1 after PLEX treatment and 0.4 for standard care ($P<.001$).[9,10,12,13] Similarly, IVIg is recommended to begin within 2 to 4 weeks of neuropathy onset in patients requiring aid to walk. This recommendation is supported by meta-analysis[14] of 2 trials comparing IVIg versus PLEX.[15,16] Because PLEX efficacy in AIDP had been determined before crafting the IVIg trials, comparison of IVIg to placebo does not exist. However, in patients with AIDP, median time until recovery of unaided ambulation was the same (approximately 50 days) in one trial[15] and shorter in the IVIg-treated group compared with PLEX treatment (55 vs 69 days, $P<.05$) in the second trial.[16] Nonetheless, with a z-score of 0.61 ($P = .54$) for recovery of ambulation and nonsignificant difference in disability grade at 4 weeks after symptom onset of 0.11 (95% CI = −0.14–0.37) in meta-analysis,[14] PLEX and IVIg are comparably efficacious, even in advanced disease. However, PLEX may represent greater risks and difficulty of administration. Administering both treatments has not proved to be superior to administering either treatment alone, and there is concern that PLEX may interfere with IVIg efficacy by removing the antibody. Corticosteroids, once thought to be beneficial for patients with AIDP, no longer play a role in treatment, because significant benefits in recovery have not been found.[17]

Because of the different immunologic causes of various subtypes of AIDP, which are beyond the scope of this review, the use of certain therapies that target the immune system have been proposed. Some of these include immunoabsorption, medications that inhibit the complement cascade, medications that inhibit T cells, and interferon-β1a. Insufficient evidence on efficacy of these therapies is available at this time.[7]

Airway Management

Although prognosis for AIDP is usually good, up to 25% of patients affected by AIDP will require mechanical ventilation for respiratory failure. Thus, respiratory function should be closely monitored even in mild AIDP, because respiratory muscles weakness and bulbar dysfunction may develop rapidly as the disease progresses before eventual improvement. Impending respiratory failure may be predicted by a vital capacity less than 20 mL/kg, maximum inspiratory pressure (MIP) less than 30 cm H_2O, and MIP less negative than -40 cm H_2O. In intubated patients, tracheostomy should be considered if pulmonary function tests have not improved from baseline after several days of mechanical ventilation. With improvement in pulmonary function, the need for tracheostomy should be reassessed periodically before extubation.[18] Noninvasive positive pressure ventilation is typically not useful in patients with AIDP due to longevity of impairment and, in the setting of bulbar dysfunction, may be contraindicated.[19] Notably, bulbar dysfunction can occur in isolation from respiratory muscle weakness; thus, endotracheal intubation may be required for airway protection, despite satisfactory pulmonary function studies. In the setting of severe bulbar weakness, gastrostomy feeding tubes may be required if oral intake is unsafe. In fact, a variety of AIDP variants exist that preferentially affect different muscles groups, and patients may have profound bulbar weakness with only modest extremity weakness, or vice versa.

Pain Management

Pain, thought to be predominantly neuropathic, has been reported in up to 89% of patients with AIDP.[20] However, systematic review on pharmacologic interventions for pain in AIDP found insufficient evidence to support use of any specific pharmacologic agent or agents.[20] Two small, randomized control trials (RCTs) found improvement in pain scales (1–10 scale) with gabapentin in the acute phase of AIDP.[21,22] Mean pain scores improved from 7.2 ± 0.83 (day 0) to 2.06 ± 0.63 (day 7) after treatment with gabapentin but did not improve after placebo (7.83 ± 0.78 [day 0] to 5.67 ± 0.91 [day 7]).[21] Compared with placebo, carbamazepine treatment after AIDP does not decrease median pain scores between day 1 and day 3 (3.0 vs 4.0 for each day) of treatment but is effective between day 4 and day 7 (carbamazepine vs placebo: 4.0 vs 6.0 on day 4 to –3.0 vs 6.0 on day 7; $P<.05$).[22] A large RCT did not reveal an improvement in pain scores with a 5-day course of methylprednisolone.[23] Acetaminophen and nonsteroidal anti-inflammatory drugs may be used in the management of pain in patients with AIDP, but these medications often fail to provide full pain relief.[18] Opioids may be useful but must be carefully considered and titrated, because these may lead to negative repercussions, especially in the setting of autonomic or respiratory dysfunction. Adjuvant therapy for long-term pain management may also include tricyclic antidepressants, tramadol, gabapentin, carbamazepine, and mexiletine.[18]

Autonomic Dysfunction Management

Autonomic dysfunction occurs in approximately two-thirds of patients with AIDP. Symptoms may include bradyarrhythmias, hypertension, hypotension, bladder and bowel dysfunction, and pupillary and sweating abnormalities.[18] These symptoms more commonly affect patients with advanced generalized weakness and respiratory failure. Blood pressure and heart rate should be closely monitored until ventilatory support is no longer required or until patients have begun to recover without needing ventilatory support.[18] Pharmacologic interventions or transcutaneous cardiac pacing may be needed in the setting of recurrent or refractory arrhythmias. Blood pressure should be supported with fluid resuscitation to maintain normovolemia and vasoactive agents in the setting of refractory hypotension. Similarly, hypertension may require treatment with scheduled antihypertensives or continuous parenteral infusions in extreme cases or with large, frequent fluctuations. Bowel sounds and passage of stool should be evaluated daily for possible paralytic ileus. If interruption or discontinuation of enteral feeding is required, erythromycin and neostigmine may be useful in the setting of ileus, whereas promotility agents should be approached with caution to avoid possible bowel perforation.[18] Finally, patients should be monitored for urinary retention and bladder catheters placed, or intermittent catheterization should occur to avoid discomfort and bladder rupture.

Rehabilitation

Evidence-based recommendations on long-term rehabilitation for patients who have endured AIDP are currently lacking.[24] However, physical (PT) and occupational therapy (OT) are obvious critical elements in recovery after AIDP. Factors to consider while planning for AIDP rehabilitation include the possibility of fatigue, muscle shortening, joint contractures, postural hypotension, weight loss, and decubitus ulcers. Rehabilitation should focus on muscle-strengthening exercises during the acute phase, and proper limb positioning, posture, orthotics, and nutrition thereafter.[18] Finally, because severe cases of AIDP may require prolonged ICU stays, superimposed critical illness myopathy (CIM) and polyneuropathy (CIP) may occur and should be factored into rehabilitation planning (see later discussion).

Conclusion

AIDP is a rapidly progressive peripheral neuropathy that may affect multiple organ systems. Although prognosis is usually good, this syndrome may result in life-threatening complications early in its course. Early treatments include immunotherapy with IVIg or PLEX. Supportive treatment is also essential, including airway, pain, and autonomic dysfunction management. Residual deficits after the acute phase may be life altering, for which rehabilitation should be sought.

AIDP should be distinguished from CIDP by clinical history, presentation, and course. CIDP may be related to AIDP and, in many ways, might be considered its chronic counterpart. Presenting with sensory neuropathy and progressive weakness, CIDP is most common in young men. As in AIDP, treatment of CIDP may include PLEX and IVIg, but, in contrast to AIDP, may also include steroids and immunosuppressants, and the prognosis varies widely.

AMYOTROPHIC LATERAL SCLEROSIS
Overview

Amyotrophic lateral sclerosis (ALS) is a progressive, incurable, and fatal, yet temporarily manageable, neurodegenerative disease that affects both upper and lower motor neurons (LMNs). Upper motor neuron (UMN) dysfunction is characterized by spasticity, weakness, and brisk deep tendon reflexes. LMN dysfunction is characterized by fasciculations, muscle wasting, and weakness.[25] Different varieties of ALS exist, broadly classified as limb-onset ALS, featuring a combination of UMN and LMN signs; bulbar-onset ALS, characterized by early dysarthria and dysphagia with subsequent UMN and LMN signs after disease progression; primary lateral sclerosis, with UMN involvement only; or progressive muscular atrophy, with LMN involvement only.[25] Frontal lobe cognitive impairment and frontotemporal dementia may often be associated with ALS.[26,27] Half of ALS patients succumb within

30 months of the onset of symptoms, although approximately 20% survive between 5 and 10 years after the onset of symptoms.[25] Factors associated with shorter survival time are older age at symptom onset, early respiratory muscle dysfunction, and bulbar-onset disease.[25] In contrast, independent predictors of prolonged survival include younger age at the onset of symptoms, longer diagnostic delay, and limb-onset disease.[25]

The pathogenesis of ALS appears multifactorial and complex, and the exact mechanisms of abnormality are still unknown. Although glutamate-mediated excitotoxicity is emerging as a relevant molecular mechanism leading to ALS, multiple factors, genetic and molecular, have been implicated. Some of the most relevant gene mutations that could play an important role in the pathogenesis of ALS include mutations in chromosome 9 open reading frame 72 (c9orf72), TAR DNA-binding protein 43 (TDP-43), and fused in sarcoma (FUS) that dysregulate RNA metabolism, which leads to intracellular aggregates formation.[28] In addition, mutations of the copper/zinc superoxide dismutase-1 (SOD1) gene may lead to neurodegeneration through multiple proposed mechanisms, including increased oxidative stress, intracellular aggregates formation, neurofilament accumulation, and axonal transport dysfunction.[28]

Pharmacologic Treatment

Advances in knowledge of the cause of ALS may lead to advances in treatment in the near future. Evidence supporting the use of stem cells transplantation, gene expression modulation, and autoimmune strategies has been flourishing.[28] One such promising advancement is the development of ISIS 333611 (also referred to as ISIS-SOD1-Rx), an antisense oligonucleotide that targets messenger RNA (mRNA) in a sequence-specific manner. ISIS 333611 causes degradation of mRNA by activating the nuclear enzyme RNase H, decreasing the concentration of SOD1 mRNA and protein in the cerebrospinal fluid of rats.[29] During a phase I trial, this drug was administered as an intrathecal infusion to individuals with SOD1-positive familial ALS and was found to be well tolerated.[30] Although further investigations are ongoing, controversy exists over the clinical relevance of this drug, because only 2% of all ALS cases (20% of familial cases) are caused by SOD1 mutations.[25]

Another therapeutic advance in ALS is the development of ozenazumab, an anti-Nogo A antibody that promotes axonal regeneration.[28] Nogo A is a neurite outgrowth inhibitor isoform that is upregulated in ALS and may be a biomarker of the disease.[31] Human trials have deemed this drug well tolerated when administered intravenously, and studies assessing its pharmacokinetic and pharmacodynamic effects are ongoing.[31] Despite a wealth of rapidly growing research, riluzole, an antiglutaminergic agent, is the only disease-modifying drug available for ALS at this time.[26,27] A systematic review examining riluzole and the survival of patients with ALS concluded that riluzole probably prolongs median survival by approximately 2 to 3 months.[32] It is considered a safe drug, but it should be discontinued if liver function tests exceed 5 times the upper level of normal or if neutropenia develops.[26]

Airway Management

The most common cause of death from ALS is respiratory failure. Forced vital capacity (FVC) is the most commonly used pulmonary measurement to determine impending respiratory failure in ALS. However, FVC is not an overly sensitive predictor, because patients with FVC greater than 70% of predicted value may have respiratory failure, as evidenced by MIP less negative than -60 cm H_2O. An FVC cutoff of 75% may be more appropriate for monitoring of respiratory function and the need for ventilatory assistance.[26] In addition, sniff nasal pressures and sniff transdiaphragmatic pressures have been suggested as possible superior alternatives to FVC, given strong correlation with nocturnal oximetry and carbon dioxide measurements, respectively, as well as apnea-hypopnea index in polysomnography, in patients without bulbar dysfunction.

Both invasive and noninvasive ventilatory support play significant roles in the treatment of respiratory failure in patients with ALS. Noninvasive ventilation typically consists of bilevel positive airway pressure (BPAP). In an RCT, investigators found that ALS patients without severe bulbar dysfunction treated with noninvasive ventilation had a median survival benefit of 205 days and improved quality of life compared with those who underwent standard care.[33]

Ventilation via tracheostomy should be offered as an alternative to patients who do not tolerate noninvasive ventilation. This option should also be presented to patients who tolerate noninvasive ventilation during early discussions about goals of care, because this may be an alternative in the future, given that BPAP will eventually become ineffective as the disease progresses. Tracheostomy also prolongs survival of ALS patients (10.39 months mean survival for patients who agree to tracheostomy vs 0.83 months for patients who refuse tracheostomy; $P<.0001$ in prospective study).[34] However, most patients cannot afford the costs associated with such protracted mechanical ventilation.[26] In addition, although quality of life with tracheostomy may be comparable to quality of life with noninvasive ventilation for patients with ALS, caretakers have rated their own quality of life lower than that of the patients' after tracheostomy. Once a tracheostomy is in place, the most common cause of mortality in ALS patients is acute respiratory tract infection.[35,36]

Diaphragmatic pacing once seemed like a promising and attractive procedure for individuals with ALS after a pilot study of 16 subjects reported increased diaphragmatic movement, increased muscle thickness, and decreased decline in FVC, without safety issues.[37] However, an RCT later rejected diaphragmatic pacing as routine treatment for patients with ALS and respiratory failure.[38] In this study, patients who were treated with diaphragmatic pacing in addition to noninvasive ventilation had decreased survival and increased number of adverse events, compared with the patients who were treated with noninvasive ventilation alone (11.0 months survival vs 22.5 months survival, respectively, $P = .006$).[38]

Patients with ALS often encounter progressive bulbar weakness leading to difficulty handling oral secretions. These issues are further exacerbated by sialorrhea and difficulty coughing because of weakness of expiratory muscles.[27] Evidence-based data for the management of oral secretions in patients with ALS are limited; thus, much of the management is based on data from non-ALS patients.[27] Medical management includes β-receptor agonists, anticholinergic bronchodilators (such as ipratropium), nebulized saline, and mucolytics, such as guaifenesin. Procedural treatments include injection of botulinum toxin[39] and radiotherapy to the salivary glands.[40] The benefits of treating sialorrhea must be weighed against the risks of increased viscosity of secretions, resulting in difficulty expectorating, mucus plugs, and atelectasis.[27]

Mechanical insufflation-exsufflation aids in clearing secretions by increasing the peak cough expiratory flows, especially in patients without bulbar disease.[41,42] In patients ventilated through a tracheostomy, mechanical insufflation-exsufflation with an inflated cuff may be more efficacious in clearing oral secretions than the traditional tracheal suctioning.[43] In an RCT, high-frequency chest wall oscillation was helpful in decreasing the symptom of breathlessness, but did not significantly change any of the measured pulmonary parameters, such as FVC.[44]

Dysphagia

As ALS progresses, dysphagia from bulbar weakness increases the patient's risk for aspiration. Gastrostomy feeding tube placement may stabilize weight and provide an alternate route for medication administration. The optimal time for this procedure is uncertain. Placement before FVC decreases to less than 50% of predicted value has been recommended to decrease the risks of the procedure. Two small retrospective studies have recently suggested that the patient's level of respiratory impairment may not correlate with increased periprocedural risks for gastrostomy tube placement. One found no statistically significant difference in the rate of complications and survival between ALS patients with FVC less than 30% than in ALS patients with FVC more than 30% of predicted value at the time of gastrostomy tube placement.[45] Another study reported no difference in the mean survival rate after percutaneous endoscopic gastrostomy tube placement between ALS patients with FVC more than 50% compared to ALS patients with FVC less than 50% of predicted value.[46] Placement of percutaneous endoscopic gastrostomy, radiologically inserted gastrostomy, and per-oral gastrostomy are all considered equally safe in patients with ALS.[47]

Conclusion

ALS is a condition of progressive UMN and LMN loss resulting in global muscle weakness that ultimately impairs respiratory and laryngeal muscles. For this reason, special attention to airway management is required. Goals of care should be discussed early after diagnosis and frequently reassessed during disease progression, as respiratory failure is the most common cause of death for ALS patients. Once the trachea has been intubated in advanced disease, patients may not be able to separate from mechanical ventilation. In fact, as the disease progresses to include cranial nerves, patients may remain in a functionally locked-in state, implicating extraordinary psychological and economic repercussions for the patient, caregivers, and loved ones.

MYASTHENIA GRAVIS
Overview

MG is an autoimmune disease that affects the neuromuscular junction at the level of the postsynaptic muscle membrane. MG is now being categorized by individual antibodies targeting different components of the postsynaptic membrane, including AChR, muscle-specific kinase (MUSK), and lipoprotein-related protein 4 (LRP4). Clinically, MG is characterized by fatigable weakness of voluntary muscles. In contrast, Lambert-Eaton myasthenic syndrome, an autoimmune disease directed at calcium channels on the presynaptic membrane of the neuromuscular junction, causes a brief increase in muscle strength with exertion, with subsequent fatigue.[48]

A diagnosis of MG is confirmed by positive serum testing for antibodies against AChR, MUSK, or LRP4 in individuals with symptoms consistent with the disease. Neurophysiologic tests, such as repetitive nerve stimulation and single-fiber electromyography, are unnecessary in individuals with the characteristic clinical symptoms and positive antibody testing. However, neurophysiologic tests become helpful in diagnosing symptomatic patients with undetectable antibodies. An edrophonium (Tensilon) test should be used only when diagnosis is urgent and in a facility with the appropriate resources for resuscitation, because of bradycardia and/or hypotension elicited by this test.[49] The test involves intravenous injection of edrophonium, a fast-onset and short-acting acetylcholinesterase inhibitor (AChE-I), to assess momentary improvement of muscle weakness. Of note, this test can be positive in subjects with other conditions and even in normal subjects.[49]

Symptomatic Treatment

Symptom alleviation for MG is most commonly achieved through chronic oral administration of pyridostigmine. Pyridostigmine is AChE-I that raises the level of acetylcholine (ACh) available in the neuromuscular junction by reducing its breakdown.[50] Doses rarely exceed 450 to 600 mg/d, as a higher dose could result in increased muscle weakness and precipitate a cholinergic crisis.[50] Interestingly, individuals with MUSK antibody-positive MG are more likely to have poor response or intolerance to AChE-I.[51]

Immunosuppressive Treatment

If symptomatic treatment does not achieve satisfactory functional status, immunosuppressive therapy is often started. A combination of immunosuppressive agents is usually preferred, in order to maximize benefits and minimize side effects. Evidence on this type of treatment is limited, for which recommendations are largely based on limited data and practitioners' experiences.

Steroids are the fastest-acting immunomodulating medications available for MG at this time.[52] Prednisolone and prednisone are usually started at high doses, such as 0.75 to 1 mg/kg/d, and titrated until optimization of symptoms is achieved. The dose should then be gradually tapered, until the lowest dose needed to remain in remission is found. Alternate-day dosing has been suggested to minimize side effects. Of note, patients with MG will often notice worsening of their symptoms with steroid initiation with gradual improvement over the first few weeks.

Long-term azathioprine, a purine analogue, in combination with short-term corticosteroids is a first-line immunosuppressive therapy. Azathioprine in combination with alternate-day prednisolone may increase remission time and decrease relapses, side effects, and prednisolone dose (at 2 and 3 years), when compared with alternate-day prednisolone alone.[53] Second-line immunosuppressive agents for the treatment of MG include methotrexate, cyclophosphamide, mycophenolate mofetil, tacrolimus, and cyclosporine. Cyclosporine and cyclophosphamide should be considered last, given their serious adverse-effects profile.[54]

Rituximab is a monoclonal antibody that binds to transmembrane antigen CD20 and leads to B-lymphocytes depletion. It has been approved for the treatment of some lymphomas and certain patients with rheumatoid arthritis.[52] Nevertheless, it has been successfully applied to the treatment of refractory MG in multiple small uncontrolled trials and case reports.[54–59] Rituximab has also been particularly beneficial in the treatment of MUSK antibody-positive MG.[55–59]

The benefits of a thymectomy in patients with MG and a thymoma are clear. However, the efficacy of thymectomy in nonthymomatous MG patients remains controversial, mainly due to the lack of prospective randomized studies.[60] Multiple controlled studies have reported increased rate of remission and improvement of symptoms, especially in patients with generalized and severe MG.[60] For this reason, it is recommended as an option for the treatment of MG.[60] Many experts do not recommend this procedure for MUSK antibody-positive MG, because efficacy has not been found and there is minimal to nonexistent thymic abnormality in these individuals.[61]

IVIg and PLEX (see **Table 2**) are usually reserved for the treatment of severe MG exacerbations and MG crises.[62] An RCT comparing IVIg and PLEX concluded these therapies are equally effective in patients with moderate to severe MG, the duration of the effect is similar, and both are well tolerated.[63] Patients positive for ACh receptor antibodies and those with more severe disease at baseline may be more responsive to therapy.[63] Another RCT found no significant difference between the efficacy of IVIg 1 g/kg and IVIg 2 g/kg in the treatment of MG exacerbation.[64] Data are currently insufficient to determine whether IVIg or PLEX is beneficial as maintenance therapy for patients with chronic MG without exacerbation.

Conclusion

MG is a complex autoimmune disease with multiple known subtypes, all of which respond differently to different treatment modalities. AChE-I, immunosuppressive drugs, and immunosuppressive therapies all play important roles in the treatment of MG. More research is needed on how and when to use these different treatment modalities on each MG subtype.

CRITICAL ILLNESS POLYNEUROPATHY AND MYOPATHY
Overview

CIP affects motor and sensory nerve axons. Clinically, CIP is most commonly characterized by patients with critical illness (eg, sepsis, multiorgan dysfunction) experiencing limb weakness or difficulty with weaning from mechanical ventilation. On neurophysiologic testing, CIP demonstrates axonal neuropathy with reduced amplitude of compound muscle action potentials (CMAPs) and sensory nerve action potentials, as well as normal conduction velocities and normal responses to repetitive nerve stimulation.

CIM affects muscles directly, not through muscle denervation. Clinically, it presents similarly to CIP, but sensory function remains intact. Therefore, as expected, amplitude of CMAPs is reduced, whereas sensory nerve action potentials remain normal. CMAP duration is increased, unlike in CIP. Muscle biopsy, although invasive and often impractical, might aid in distinguishing CIM from other myopathies, if necessary. The most common histologic finding in CIM is thick myosin filament loss. Given all the similarities between CIP and CIM, the limitations of neurophysiologic testing that are beyond the scope of this review, and the fact that these 2 diseases may often coexist, differentiating between CIP and CIM may sometimes not be possible.[65] Furthermore, rapidly advancing neuromuscular disease of any cause, warranting admission to the ICU on presentation, may be confused with CIP/CIM, especially when patients fail to extubate. Finally, weakness related to a patient's underlying cause of critical illness may be confused as or exacerbated by CIP/CIM.

Treatment

Many strategies have been proposed for the management of CIP and CIM, but no specific regimen has emerged as a definitive treatment strategy. For this reason, the management is mainly preventative and supportive.

Several reports and studies have suggested a causal relationship between NMBA use and ICU-acquired weakness.[66] A causal relationship is plausible, because NMBAs have been associated with functional denervation of the muscle, presumably causing denervation atrophy of the muscle. Furthermore, increased capillary permeability caused by sepsis may allow a direct toxic effect of NMBAs on peripheral nerves.[67] For this reason, conservative use of these drugs has been advocated as a preventative measure for many years. However, study designs addressing this question contain multiple confounding factors, such as the concomitant administration of corticosteroids, sedation, and greater severity of illness in those who developed ICU-acquired weakness after the administration of an NMBA.[66]

Steroids have faced a similar history in this regard. Prospective studies have been unable to definitively link corticosteroid use to CIP or CIM.[68,69] An RCT assessing the use of corticosteroids in persistent acute respiratory distress syndrome did not find a significant difference in the rate of clinically suspected neuromyopathy.[70] The investigators suggest that perhaps corticosteroids increase the severity but not the incidence of neuromyopathy, given that all 9 reports of serious adverse effects related to neuropathy or myopathy in this trial were in the corticosteroid group.[70] On the other hand, the absence of steroid treatment during ICU hospitalization had the strongest positive correlation with longer ambulation distances greater than 6 minutes at 3 months after discharge. In multivariate regression analysis ($R^2 = 0.31$) from a prospective study, lack of steroid administration achieved the highest β coefficient (127.0 ± 47.5) with the lowest P value (.009) of all the independent variables studied.[71] Notably, independent variables studied included age, female sex, total days in the ICU, slope of Lung Injury Score, and APACHE II (Acute Physiology and Chronic Health Evaluation) score.[71] Similarly, a separate prospective cohort reported the administration of corticosteroids as an independent predictor of the development of ICU-acquired paresis.[72]

Although conservative use of NMBAs and steroids (reduced doses and duration of administration) in critically ill patients may be prudent, and perhaps even intuitive, cause-and-effect relationships have not been definitively established between NMBAs, steroids, CIP, or CIM.

Intensive insulin therapy has been validated as a method for preventing CIP and CIM by 2 RCTs.[73,74] After multivariate logistic regression analysis and correction for known risk factors, blood glucose control was an independent protective factor against the development of CIP in the intensive insulin therapy group (odds ratio [OR] 1.26, 95% CI 1.09–1.46 per millimole blood glucose, $P = .002$).[73] Likewise, in a separate multivariate logistic regression analysis, intensive insulin therapy was an independent protective factor against a diagnosis of CIP/CIM (OR 0.61, $P = .02$, 95%CI 0.43–0.92).[74] Although intensive insulin therapy did increase the risk of hypoglycemia in these studies, the risk of death within 24 hours of the hypoglycemic events did not increase.[73,74]

The benefits of PT/OT in patients who have been critically ill are documented.[75–78] However, an RCT evaluating outcomes after early (median 1.5 days after intubation in the intervention group vs 7.4 days after intubation in the control group) PT/OT in sedated ICU patients requiring mechanical ventilation did not find a significant difference in the incidence of ICU-acquired paresis at hospital discharge.[79] Evidence is currently insufficient to determine whether or when PT/OT may be effective in preventing CIP/CIM. Nevertheless, PT/OT should be considered as soon as possible in the critically ill, because early intervention may be beneficial in other ways, such as increasing the likelihood of return to independent functional status at hospital discharge, shorter duration of delirium, and more ventilator-free days.[79]

Electrical muscle stimulation throughout the ICU admission may be beneficial in the prevention of CIP/CIM, although this topic requires further research.[80]

Finally, IVIg, nutritional interventions, antioxidant therapy, testosterone, and growth hormone have all been proposed as interventions for the treatment and prevention of CIP/CIM, but data are currently limited.[65]

SUMMARY

CIP and CIM are conditions of significant morbidity, including prolonged mechanical ventilation. Modification of risk factors, such as systemic inflammatory response syndrome and hyperglycemia, is of utmost importance. PT and OT are important in the return of functional status after discharge from the hospital. Other therapies require further study.

REFERENCES

1. Romero A, Joshi GP. Neuromuscular disease and anesthesia. Muscle Nerve 2013;48(3):451–60.
2. Blichfeldt-Lauridsen L, Hansen BD. Anesthesia and myasthenia gravis. Acta Anaesthesiol Scand 2012;56(1):17–22.
3. Wipfli M, Arnold M, Luginbuhl M. Repeated spinal anesthesia in a tetraparetic patient with Guillain-Barre syndrome. J Clin Anesth 2013;25(5):409–12.
4. Kochi T, Oka T, Mizuguchi T. Epidural anesthesia for patients with amyotrophic lateral sclerosis. Anesth Analg 1989;68(3):410–2.
5. Almeida C, Coutinho E, Moreira D, et al. Myasthenia gravis and pregnancy: anaesthetic management–a series of cases. Eur J Anaesthesiol 2010;27(11): 985–90.
6. Prabhakar A, Owen CP, Kaye AD. Anesthetic management of the patient with amyotrophic lateral sclerosis. J Anesth 2013;27(6):909–18.
7. Hughes RA, Cornblath DR. Guillain-Barre syndrome. Lancet 2005;366(9497): 1653–66.
8. Raphael JC, Chevret S, Hughes RA, et al. Plasma exchange for Guillain-Barre syndrome. Cochrane Database Syst Rev 2001;(2):CD001798.
9. Greenwood RJ, Newsom-Davis J, Hughes RA, et al. Controlled trial of plasma exchange in acute inflammatory polyradiculoneuropathy. Lancet 1984;1(8382): 877–9.
10. Plasmapheresis and acute Guillain-Barre syndrome. The Guillain-Barre syndrome Study Group. Neurology 1985;35(8):1096–104.
11. Osterman PO, Fagius J, Lundemo G, et al. Beneficial effects of plasma exchange in acute inflammatory polyradiculoneuropathy. Lancet 1984;2(8415):1296–9.
12. Efficiency of plasma exchange in Guillain-Barre syndrome: role of replacement fluids. French Cooperative Group on Plasma Exchange in Guillain-Barre syndrome. Ann Neurol 1987;22(6):753–61.
13. Appropriate number of plasma exchanges in Guillain-Barre syndrome. The French Cooperative Group on Plasma Exchange in Guillain-Barre Syndrome. Ann Neurol 1997;41(3):298–306.
14. Hughes RA, Raphaël JC, Swan AV, et al. Intravenous immunoglobulin for Guillain-Barre syndrome. Cochrane Database Syst Rev 2001;(2):CD002063.
15. Randomised trial of plasma exchange, intravenous immunoglobulin, and combined treatments in Guillain-Barre syndrome. Plasma Exchange/Sandoglobulin Guillain-Barre Syndrome Trial Group. Lancet 1997;349(9047):225–30.
16. van der Meche FG, Schmitz PI. A randomized trial comparing intravenous immune globulin and plasma exchange in Guillain-Barre syndrome. Dutch Guillain-Barre Study Group. N Engl J Med 1992;326(17):1123–9.
17. Hughes RA, van Doorn PA. Corticosteroids for Guillain-Barre syndrome. Cochrane Database Syst Rev 2012;(8):CD001446.
18. Hughes RA, Wijdicks EF, Benson E, et al. Supportive care for patients with Guillain-Barre syndrome. Arch Neurol 2005;62(8):1194–8.
19. Hill NS. Neuromuscular disease in respiratory and critical care medicine. Respir Care 2006;51(9):1065–71.

20. Liu J, Wang LN, McNicol ED. Pharmacological treatment for pain in Guillain-Barre syndrome. Cochrane Database Syst Rev 2015;(4):CD009950.
21. Pandey CK, Bose N, Garg G, et al. Gabapentin for the treatment of pain in Guillain-Barre syndrome: a double-blinded, placebo-controlled, crossover study. Anesth Analg 2002;95(6):1719–23, table of contents.
22. Pandey CK, Raza M, Tripathi M, et al. The comparative evaluation of gabapentin and carbamazepine for pain management in Guillain-Barre syndrome patients in the intensive care unit. Anesth Analg 2005;101(1):220–5, table of contents.
23. Ruts L, van Koningsveld R, Jacobs BC, et al. Determination of pain and response to methylprednisolone in Guillain-Barre syndrome. J Neurol 2007;254(10): 1318–22.
24. Meythaler JM. Rehabilitation of Guillain-Barre syndrome. Arch Phys Med Rehabil 1997;78(8):872–9.
25. Kiernan MC, Vucic S, Cheah BC, et al. Amyotrophic lateral sclerosis. Lancet 2011;377(9769):942–55.
26. Radunovic A, Mitsumoto H, Leigh PN. Clinical care of patients with amyotrophic lateral sclerosis. Lancet Neurol 2007;6(10):913–25.
27. Jenkins TM, Hollinger H, McDermott CJ. The evidence for symptomatic treatments in amyotrophic lateral sclerosis. Curr Opin Neurol 2014;27(5):524–31.
28. Vucic S, Rothstein JD, Kiernan MC. Advances in treating amyotrophic lateral sclerosis: insights from pathophysiological studies. Trends Neurosci 2014;37(8): 433–42.
29. Smith RA, Miller TM, Yamanaka K, et al. Antisense oligonucleotide therapy for neurodegenerative disease. J Clin Invest 2006;116(8):2290–6.
30. Miller TM, Pestronk A, David W, et al. An antisense oligonucleotide against SOD1 delivered intrathecally for patients with SOD1 familial amyotrophic lateral sclerosis: a phase 1, randomised, first-in-man study. Lancet Neurol 2013;12(5): 435–42.
31. Meininger V, Pradat PF, Corse A, et al. Safety, pharmacokinetic, and functional effects of the nogo—a monoclonal antibody in amyotrophic lateral sclerosis: a randomized, first-in-human clinical trial. PLoS One 2014;9(5):e97803.
32. Miller RG, Mitchell JD, Moore DH. Riluzole for amyotrophic lateral sclerosis (ALS)/ motor neuron disease (MND). Cochrane Database Syst Rev 2012;(3):CD001447.
33. Bourke SC, Tomlinson M, Williams TL, et al. Effects of non-invasive ventilation on survival and quality of life in patients with amyotrophic lateral sclerosis: a randomised controlled trial. Lancet Neurol 2006;5(2):140–7.
34. Sancho J, Servera E, Díaz JL, et al. Home tracheotomy mechanical ventilation in patients with amyotrophic lateral sclerosis: causes, complications and 1-year survival. Thorax 2011;66(11):948–52.
35. Chio A, Calvo A, Ghiglione P, et al. Tracheostomy in amyotrophic lateral sclerosis: a 10-year population-based study in Italy. J Neurol Neurosurg Psychiatry 2010; 81(10):1141–3.
36. Vianello A, Arcaro G, Palmieri A, et al. Survival and quality of life after tracheostomy for acute respiratory failure in patients with amyotrophic lateral sclerosis. J Crit Care 2011;26(3):329.e7–14.
37. Onders RP, Elmo M, Kaplan C, et al. Final analysis of the pilot trial of diaphragm pacing in amyotrophic lateral sclerosis with long-term follow-up: diaphragm pacing positively affects diaphragm respiration. Am J Surg 2014;207(3):393–7 [discussion: 397].
38. DiPALS Writing Committee, DiPALS Study Group Collaborators, McDermott CJ, Bradburn MJ, Maguire C, et al. Safety and efficacy of diaphragm pacing in

patients with respiratory insufficiency due to amyotrophic lateral sclerosis (Di-PALS): a multicentre, open-label, randomised controlled trial. Lancet Neurol 2015;14(9):883–92.

39. Squires N, Humberstone M, Wills A, et al. The use of botulinum toxin injections to manage drooling in amyotrophic lateral sclerosis/motor neurone disease: a systematic review. Dysphagia 2014;29(4):500–8.

40. Assouline A, Levy A, Abdelnour-Mallet M, et al. Radiation therapy for hypersalivation: a prospective study in 50 amyotrophic lateral sclerosis patients. Int J Radiat Oncol Biol Phys 2014;88(3):589–95.

41. Mustfa N, Aiello M, Lyall RA, et al. Cough augmentation in amyotrophic lateral sclerosis. Neurology 2003;61(9):1285–7.

42. Sancho J, Servera E, Díaz J, et al. Efficacy of mechanical insufflation-exsufflation in medically stable patients with amyotrophic lateral sclerosis. Chest 2004;125(4): 1400–5.

43. Sancho J, Servera E, Vergara P, et al. Mechanical insufflation-exsufflation vs. tracheal suctioning via tracheostomy tubes for patients with amyotrophic lateral sclerosis: a pilot study. Am J Phys Med Rehabil 2003;82(10):750–3.

44. Lange DJ, Lechtzin N, Davey C, et al. High-frequency chest wall oscillation in ALS: an exploratory randomized, controlled trial. Neurology 2006;67(6):991–7.

45. Sarfaty M, Nefussy B, Gross D, et al. Outcome of percutaneous endoscopic gastrostomy insertion in patients with amyotrophic lateral sclerosis in relation to respiratory dysfunction. Amyotroph Lateral Scler Frontotemporal Degener 2013; 14(7–8):528–32.

46. Czell D, Bauer M, Binek J, et al. Outcomes of percutaneous endoscopic gastrostomy tube insertion in respiratory impaired amyotrophic lateral sclerosis patients under noninvasive ventilation. Respir Care 2013;58(5):838–44.

47. ProGas Study Group. Gastrostomy in patients with amyotrophic lateral sclerosis (ProGas): a prospective cohort study. Lancet Neurol 2015;14(7):702–9.

48. Hulsbrink R, Hashemolhosseini S. Lambert-Eaton myasthenic syndrome—diagnosis, pathogenesis and therapy. Clin Neurophysiol 2014;125(12):2328–36.

49. Thanvi BR, Lo TC. Update on myasthenia gravis. Postgrad Med J 2004;80(950): 690–700.

50. Maggi L, Mantegazza R. Treatment of myasthenia gravis: focus on pyridostigmine. Clin Drug Investig 2011;31(10):691–701.

51. Guptill JT, Sanders DB, Evoli A. Anti-MuSK antibody myasthenia gravis: clinical findings and response to treatment in two large cohorts. Muscle Nerve 2011; 44(1):36–40.

52. Sieb JP. Myasthenia gravis: an update for the clinician. Clin Exp Immunol 2014; 175(3):408–18.

53. Palace J, Newsom-Davis J, Lecky B. A randomized double-blind trial of prednisolone alone or with azathioprine in myasthenia gravis. Myasthenia Gravis Study Group. Neurology 1998;50(6):1778–83.

54. Sathasivam S. Steroids and immunosuppressant drugs in myasthenia gravis. Nat Clin Pract Neurol 2008;4(6):317–27.

55. Evoli A, Padua L. Diagnosis and therapy of myasthenia gravis with antibodies to muscle-specific kinase. Autoimmun Rev 2013;12(9):931–5.

56. Yi JS, Decroos EC, Sanders DB, et al. Prolonged B-cell depletion in MuSK myasthenia gravis following rituximab treatment. Muscle Nerve 2013;48(6):992–3.

57. Blum S, Gillis D, Brown H, et al. Use and monitoring of low dose rituximab in myasthenia gravis. J Neurol Neurosurg Psychiatry 2011;82(6):659–63.

58. Keung B, Robeson KR, DiCapua DB, et al. Long-term benefit of rituximab in MuSK autoantibody myasthenia gravis patients. J Neurol Neurosurg Psychiatry 2013;84(12):1407–9.

59. Diaz-Manera J, Martínez-Hernández E, Querol L, et al. Long-lasting treatment effect of rituximab in MuSK myasthenia. Neurology 2012;78(3):189–93.

60. Gronseth GS, Barohn RJ. Practice parameter: thymectomy for autoimmune myasthenia gravis (an evidence-based review): report of the quality standards subcommittee of the American Academy of Neurology. Neurology 2000;55(1):7–15.

61. El-Salem K, Yassin A, Al-Hayk K, et al. Treatment of MuSK-associated myasthenia gravis. Curr Treat Options Neurol 2014;16(4):283.

62. Zinman L, Ng E, Bril V. IV immunoglobulin in patients with myasthenia gravis: a randomized controlled trial. Neurology 2007;68(11):837–41.

63. Barth D, Nabavi Nouri M, Ng E, et al. Comparison of IVIg and PLEX in patients with myasthenia gravis. Neurology 2011;76(23):2017–23.

64. Gajdos P, Tranchant C, Clair B, et al. Treatment of myasthenia gravis exacerbation with intravenous immunoglobulin: a randomized double-blind clinical trial. Arch Neurol 2005;62(11):1689–93.

65. Apostolakis E, Papakonstantinou NA, Baikoussis NG, et al. Intensive care unit-related generalized neuromuscular weakness due to critical illness polyneuropathy/myopathy in critically ill patients. J Anesth 2015;29(1):112–21.

66. Puthucheary Z, Rawal J, Ratnayake G, et al. Neuromuscular blockade and skeletal muscle weakness in critically ill patients: time to rethink the evidence? Am J Respir Crit Care Med 2012;185(9):911–7.

67. Bolton CF. Neuromuscular manifestations of critical illness. Muscle Nerve 2005; 32(2):140–63.

68. de Letter MA, Schmitz PI, Visser LH, et al. Risk factors for the development of polyneuropathy and myopathy in critically ill patients. Crit Care Med 2001;29(12): 2281–6.

69. Nanas S, Kritikos K, Angelopoulos E, et al. Predisposing factors for critical illness polyneuromyopathy in a multidisciplinary intensive care unit. Acta Neurol Scand 2008;118(3):175–81.

70. Steinberg KP, Hudson LD, Goodman RB, et al. Efficacy and safety of corticosteroids for persistent acute respiratory distress syndrome. N Engl J Med 2006; 354(16):1671–84.

71. Herridge MS, Cheung AM, Tansey CM, et al. One-year outcomes in survivors of the acute respiratory distress syndrome. N Engl J Med 2003;348(8):683–93.

72. De Jonghe B, Sharshar T, Lefaucheur JP, et al. Paresis acquired in the intensive care unit: a prospective multicenter study. JAMA 2002;288(22):2859–67.

73. Van den Berghe G, Schoonheydt K, Becx P, et al. Insulin therapy protects the central and peripheral nervous system of intensive care patients. Neurology 2005;64(8):1348–53.

74. Hermans G, Wilmer A, Meersseman W, et al. Impact of intensive insulin therapy on neuromuscular complications and ventilator dependency in the medical intensive care unit. Am J Respir Crit Care Med 2007;175(5):480–9.

75. Novak P, Vidmar G, Kuret Z, et al. Rehabilitation of critical illness polyneuropathy and myopathy patients: an observational study. Int J Rehabil Res 2011;34(4): 336–42.

76. Fan E. Critical illness neuromyopathy and the role of physical therapy and rehabilitation in critically ill patients. Respir Care 2012;57(6):933–44 [discussion: 944–6].

77. Nordon-Craft A, Schenkman M, Ridgeway K, et al. Physical therapy management and patient outcomes following ICU-acquired weakness: a case series. J Neurol Phys Ther 2011;35(3):133–40.
78. Burtin C, Clerckx B, Robbeets C, et al. Early exercise in critically ill patients enhances short-term functional recovery. Crit Care Med 2009;37(9):2499–505.
79. Schweickert WD, Pohlman MC, Pohlman AS, et al. Early physical and occupational therapy in mechanically ventilated, critically ill patients: a randomised controlled trial. Lancet 2009;373(9678):1874–82.
80. Routsi C, Gerovasili V, Vasileiadis I, et al. Electrical muscle stimulation prevents critical illness polyneuromyopathy: a randomized parallel intervention trial. Crit Care 2010;14(2):R74.
81. Lunemann JD, Nimmerjahn F, Dalakas MC. Intravenous immunoglobulin in neurology–mode of action and clinical efficacy. Nat Rev Neurol 2015;11(2):80–9.
82. Lehmann HC, Hartung HP, Meyer Zu Hörste G, et al. Plasma exchange in immune-mediated neuropathies. Curr Opin Neurol 2008;21(5):547–54.
83. Reeves HM, Winters JL. The mechanisms of action of plasma exchange. Br J Haematol 2014;164(3):342–51.
84. Ebadi H, Barth D, Bril V. Safety of plasma exchange therapy in patients with myasthenia gravis. Muscle Nerve 2013;47(4):510–4.
85. Winters JL. Plasma exchange: concepts, mechanisms, and an overview of the American Society for Apheresis guidelines. Hematology Am Soc Hematol Educ Program 2012;2012:7–12.
86. Hughes RA. Give or take? Intravenous immunoglobulin or plasma exchange for Guillain-Barre syndrome. Crit Care 2011;15(4):174.

Index

Note: Page numbers of article titles are in **boldface** type.

A

Acetaminophen, for neurosurgery in patients with chronic pain, 485–487
Acidotoxicity, in ischemic brain injury, 454–455
Acute inflammatory demyelinating polyneuropathy, in the neurointensive care unit, 604–608
    airway management, 607
    autonomous dysfunction management, 607–608
    immunotherapy, 604–606
    overview, 604
    pain management, 607
    rehabilitation, 608
Acute ischemic stroke. *See* Ischemic stroke
Aging brain, vulnerability to anesthesia-induced neurotoxicity, 448
Amyotrophic lateral sclerosis, in the neurointensive care unit, 608–611
    airway management, 609–610
    dysphagia, 610
    overview, 608–609
    pharmacologic treatment, 609
Anesthetic methods, in endovascular approaches to acute ischemic stroke, **497–509**
    for patients with chronic pain undergoing neurosurgery, **479–495**
Anesthetics, general. *See* General anesthetics.
Aneurysm obliteration, in subarachnoid hemorrhage, 587–589
    coiling, 587–589
        balloon assisted, 588
        flow diversion, 588
        stent assisted, 588
    endovascular evolution, 587
Animal behavior, effects of excessive modulation of neurotransmitters during brain
    development on, 442–446
Apoptosis, in ischemic brain injury, 456–457
Autoregulation, of cerebral blood flow, **465–477**
    clinical applications of testing of, 467
    effects of anesthetics on, 467
    mechanism of, 466
    methods for assessment of, 466–467
    pathophysiology in TBI, 467–468
    role of age and sex in, and treatment of dysregulation after TBI, 468–473
    dysfunctional cerebral, in subarachnoid hemorrhage, 594
    monitoring cerebrovascular reactivity, 515–516
        other autoregulatory indices, 516
        pressure reactivity index, 516
        treatment guided by, 516
    in pathophysiology of TBI, 558–559

Anesthesiology Clin 34 (2016) 621–631
http://dx.doi.org/10.1016/S1932-2275(16)30061-1
1932-2275/16/$ – see front matter

Moving?

Make sure your subscription moves with you!

To notify us of your new address, find your **Clinics Account Number** (located on your mailing label above your name), and contact customer service at:

Email: journalscustomerservice-usa@elsevier.com

800-654-2452 (subscribers in the U.S. & Canada)
314-447-8871 (subscribers outside of the U.S. & Canada)

Fax number: 314-447-8029

Elsevier Health Sciences Division
Subscription Customer Service
3251 Riverport Lane
Maryland Heights, MO 63043

*To ensure uninterrupted delivery of your subscription, please notify us at least 4 weeks in advance of move.

Printed and bound by CPI Group (UK) Ltd, Croydon, CR0 4YY

03/10/2024

01040393-0003